United States
Environmental Protection
Agency

EPA 231K13001
June 2013
www.epa.gov/smartgrowth

Our Built and Natural Environments:

A Technical Review of the Interactions Among Land Use, Transportation, and Environmental Quality

SECOND EDITION

Office of Sustainable Communities
Smart Growth Program

Acknowledgments

The U.S. Environmental Protection Agency (EPA), through its Office of Sustainable Communities managed the preparation of this report. This report has been subjected to the Agency's peer and administrative review and has been approved for publication as an EPA document. Mention of trade names or commercial products does not constitute endorsement or recommendation for use.

Principal author: Melissa G. Kramer, Ph.D.

Contributors and reviewers from the U.S. Environmental Protection Agency:

* Office of Sustainable Communities: Danielle Arigoni, Ted Cochin, John Frece, Susan Gitlin, Jeff Jamawat, Adam Klinger, Megan McConville, Kevin Nelson, Kevin Ramsey, Megan Susman, John Thomas, Tim Torma, Brett Van Akkeren, and Beth Zgoda
* Office of Air and Radiation: Ken Adler, Stacy Angel, Chad Bailey, Laura Berry, Gregory Brunner, James Hemby, Rudy Kapichak, Jean Lupinacci, Neelam Patel, Meg Patulski, Karl Pepple, Tamara Saltman, Erika Sasser, Kimber Scavo, and Mark Simons
* Office of Chemical Safety and Pollution Prevention: Matt Bogoshian, David Sarokin, and Tom Simons
* Office of Environmental Justice: Suzi Ruhl
* Office of Policy: Alex Barron
* Office of Research and Development: Fran Kremer, Melissa McCullough, Joseph McDonald, and Barbara Walton
* Office of Solid Waste and Emergency Response: Ksenija Janjic and Patricia Overmeyer
* Office of Water: Laura Bachle, Veronica Blette, Robert Goo, Rachel Herbert, Christopher Kloss, and Jennifer Linn
* Region 5, Land and Chemicals Division: Bradley Grams

Other federal government reviewers:

* Centers for Disease Control and Prevention: Lorraine Backer, Ginger Chew, Daneen Farrow-Collier, Robynn Leidig, Erin Sauber-Schatz, Arthur Wendel, and Margalit Younger
* U.S. Department of Transportation: Alexandra Tyson
* General Services Administration: Ken Sandler
* National Oceanic and Atmospheric Administration: Sarah van der Schalie

External peer reviewers:

* Alexander Felson, Yale University
* Susan Handy, University of California Davis
* John Jacob, Texas A&M University
* Kevin Krizek, University of Colorado
* Dowell Myers, University of Southern California
* Daniel Rodriguez, University of North Carolina
* Frank Southworth, Georgia Institute of Technology
* Paul Sutton, University of Denver

Cover photo credits:

* Front cover, top left: drouu via stock.xchng
* Front cover, top right: Kyle Gradinger via flickr.com
* Back cover, top right: Dan Burden via Pedestrian and Bicycle Information Center
* All others: EPA

Table of Contents

Executive Summary

Decisions about how and where we build our communities have significant impacts on the natural environment and on human health. Cities, regions, states, and the private sector need information about the environmental effects of their land use and transportation decisions to mitigate growth-related environmental impacts and to improve community quality of life and human health.

This report:

- Discusses the status of and trends in land use, development, and transportation and their environmental implications.

- Articulates the current understanding of the relationship between the built environment and the quality of air, water, land resources, habitat, and human health.

- Provides evidence that certain kinds of land use and transportation strategies can reduce the environmental and human health impacts of development.

Patterns of development, transportation infrastructure, and building location and design—the built environment—directly affect the natural environment. Development takes the place of natural ecosystems and fragments habitat. It also influences decisions people make about how to get around and determines how much people must travel to meet daily needs. These mobility and travel decisions have indirect effects on human health and the natural environment by affecting air and water pollution levels, the global climate, levels of physical activity and community engagement, and the number and severity of vehicle crashes.

Trends in land use, building patterns, and travel behavior highlight how significantly our development patterns have changed in recent decades. The size of virtually every major metropolitan area in the United States has expanded dramatically. In many places, the rate of land development has far outpaced the rate of population growth, although more recent trends in some areas suggest the pattern could be changing. As the amount of developed land has increased and more and larger homes have been built, buildings, roads, and associated impervious surfaces have grown to serve an increasingly dispersed population. As our communities changed to accommodate cars, the percentages of people taking public transit, walking, and biking declined. Projected population growth and demographic trends suggest that the need for additional development will continue to increase.

As the U.S. population has grown, we have developed land that serves important ecological functions at a significant cost to the environment. Development has destroyed, degraded, and fragmented habitat. Water quality has declined. Air quality in many areas of the country is still adversely affecting human health. The heat island effect and global climate change illustrate just how complex and far-reaching the impacts of our built environment are. Community design can make it difficult for people to get adequate physical activity, engage with neighbors, and participate in community events. It can also increase the risk of injury or death from a vehicle crash.

Changing where and how we build our communities can help mitigate these impacts, improving how development affects the environment and human health:

- **Where we build** involves locating development in a region or land area. It includes safeguarding sensitive areas such as riparian buffers, wetlands, and critical habitat from development pressures; directing new development to infill, brownfield, and greyfield sites to take advantage of existing infrastructure and preserve green space; and putting homes, workplaces, and services close to each other in convenient, accessible locations.

- **How we build** includes developing more compactly to preserve open spaces and water quality; mixing uses to reduce travel distances; designing communities and streets to promote walking and biking; and improving building design, construction, and materials selection to use natural resources more efficiently and improve buildings' environmental performance.

These elements are interrelated and often work most effectively in combination with each other rather than individually. Although findings might differ on the magnitude of the effects of different practices, the evidence is overwhelming that some types of development yield better environmental results than others. Used in combination, these practices can significantly reduce impacts on habitat, ecosystems, and watersheds and can reduce vehicle travel and energy use, which in turn reduces emissions that cause local, regional, and global air quality concerns. As communities nationwide look for ways to reduce the environmental and human health impacts of their development decisions, the evidence is clear that our nation can continue to grow and can build a strong foundation for lasting prosperity while also protecting our environment and health.

Chapter 1. Introduction

1.1 Purpose

Recognition is increasing that land use and transportation decisions can either support or interfere with environmental protection and quality of life. Policy-makers have realized that decisions about how and where we build our communities have significant impacts on the natural environment. Cities, regions, states, and the private sector are planning and implementing smart growth strategies and other measures to mitigate growth-related environmental impacts and to improve community quality of life and human health (Exhibit 1-1).

This edition of *Our Built and Natural Environments* updates the original 2001 publication with the most current information available as of October 2012.[1] It is written for everyone interested in how land use practices, transportation infrastructure, and building siting and design directly and indirectly affect environmental quality.

This report provides information that can help state and local governments decide how to accommodate expected population growth within their borders in the most environmentally responsible manner. Different parts of the country face different challenges and opportunities based on the availability of fresh water, the mix of fossil fuel and renewable energy sources, and their

> **Exhibit 1-1: Smart Growth.** Smart growth strategies create sustainable communities by siting development in convenient locations and designing it to be more efficient and environmentally responsible. Communities across the country are using creative strategies to develop in ways that preserve natural lands and critical environmental areas, protect water and air quality, and reuse already-developed land. They conserve resources by reinvesting in existing infrastructure and reclaiming historic buildings. By designing neighborhoods that have shops, offices, schools, churches, parks, and other amenities near homes, communities are giving their residents and visitors the option of walking, bicycling, taking public transportation, or driving as they go about their business. A range of different types of homes makes it possible for senior citizens to stay in their homes as they age, young people to afford their first home, and families at all stages in-between to find a safe, attractive home they can afford. Through smart growth approaches that enhance neighborhoods and involve local residents in development decisions, these communities are creating vibrant places to live, work, and play. The high quality of life in these communities makes them economically competitive, creates business opportunities, and improves the local tax base.

vulnerability to natural disasters, among other issues. Whether or to what extent growth should occur in a particular region is beyond the scope of this document.

---

[1] This paper generally uses the most recent literature. Studies older than 2005 are included only if they made an important finding that is not repeated in later research. The document cites review articles when they are available rather than individual studies to make the paper more concise. The document relies primarily on peer-reviewed academic literature, reports by the National Academy of Sciences, and government publications that report on government-collected data. Reports published by nonprofit organizations and private contractors are not included because of uncertainty about the level of peer review and potential perceived bias.

This report also does not discuss the economic impacts of different land use, development, or transportation decisions. The economic implications of various choices influence community decision-making, and rightly so. However, this document focuses on how different choices can help protect human health and the environment. EPA has published other reports that discuss some of the ways certain types of development can provide better economic outcomes for businesses, local governments, and households while also protecting the environment and improving quality of life.[2]

This report concludes that the built environment and the travel decisions it encourages have significant impacts on the environment, human health, and community quality of life. As a result, it is important for EPA; other federal, state, and local government agencies; real estate developers and investors; and communities across America to understand the relationship between the built and natural environments.

1.2 The Effects of the Built Environment on Human Health and the Natural Environment

For decades, people have recognized the environmental impacts resulting from industrial pollution. The environmental effects of land use decisions are not as widely understood in spite of their tremendous impact. Patterns of development, transportation infrastructure, and building location and design—the built environment—directly affect the natural environment. Development takes the place of natural ecosystems and fragments habitat. It also influences decisions people make about how to get around and determines how much people must travel to meet daily needs. These mobility and travel decisions have indirect effects on human health and the natural environment by affecting air and water pollution levels, the global climate, levels of physical activity and community engagement, and the number and severity of vehicle crashes (Exhibit 1-2).

1.2.1 Direct Effects

The extent of land development, the type of development, and the location of infrastructure have direct and long-lasting impacts on ecosystems. Natural ecosystems serve a variety of functions that provide people with necessary and valuable goods and services. For example, natural ecosystems maintain healthy air quality, regulate temperature and precipitation, prevent flooding, provide clean water for drinking and industrial use, maintain healthy and productive soil, pollinate wild plants and crops, maintain biological and genetic diversity, provide renewable natural resources, treat organic waste, control pests and diseases, and provide recreation areas.[3] Land development often replaces natural areas and damages or destroys many of these ecosystem functions and services. Land development also frequently affects the amount and quality of essential habitat for plants and animals.

---

[2] EPA's Office of Sustainable Communities has a variety of resources on the economic impacts of smart growth strategies. See the Business and Economic Development section of the office's publications page: *www.epa.gov/smartgrowth/publications.htm#bizdev*.
[3] de Groot, Wilson, and Boumans 2002

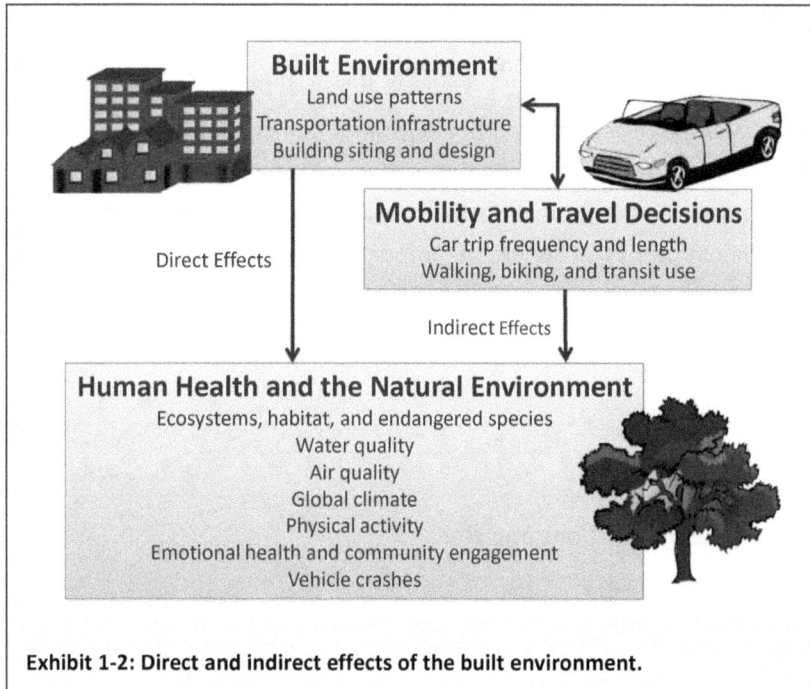

Built Environment
Land use patterns
Transportation infrastructure
Building siting and design

Mobility and Travel Decisions
Car trip frequency and length
Walking, biking, and transit use

Direct Effects

Indirect Effects

Human Health and the Natural Environment
Ecosystems, habitat, and endangered species
Water quality
Air quality
Global climate
Physical activity
Emotional health and community engagement
Vehicle crashes

Exhibit 1-2: Direct and indirect effects of the built environment.

1.2.2   Indirect Effects

How we develop land determines the distribution of jobs, housing, and community activities, which in turn determines how far people travel to meet their daily needs and their transportation options. These travel decisions affect air pollutant and greenhouse gas emissions and levels of physical activity. Land development patterns can disproportionately affect children, the elderly, and people with disabilities—groups that can be especially vulnerable to the health effects of pollution. In addition, these groups, as well as racial and ethnic minorities and people of lower socioeconomic status, often have fewer options for where to live and work.[4,5]

Travel behavior is thus one of the most important indirect effects of how and where we build and is a topic of intense investigation. Travel behavior is complex, with various factors simultaneously affecting decisions about how much, where, when, and how to get around. Although the magnitude of the effect is widely debated, a large body of evidence shows that community design affects travel behavior. Moreover, the effect is large enough that changes in community design could help mitigate a range of transportation, air quality, human health, and greenhouse gas problems.

1.3   Overview of Document

Chapter 2 covers current status of and trends in land use and travel behavior—how our population has grown; how we have developed land and constructed buildings and roads to accommodate growth; how our use of cars has grown; and how transit use, walking, and biking have changed over time.

Chapter 3 looks at how these trends in land use and transportation have affected the environment, including habitat loss, degradation, and fragmentation; degradation of water resources and water quality; degradation of air quality; and global climate change. Chapter 3 also considers the human health

---

[4] Younger, et al. 2008
[5] EPA supports *environmental justice*, the fair treatment and meaningful involvement of all people regardless of race, color, national origin, or income with respect to the development, implementation, and enforcement of environmental laws, regulations, and policies. Consideration of how our development patterns affect disadvantaged and vulnerable populations helps EPA achieve this goal. See: *www.epa.gov/environmentaljustice*.

effects of our built environment, including impacts on activity levels, obesity, chronic disease, level of community engagement, and risk of injury or death from vehicle crashes.

Chapter 4 provides evidence that some patterns of growth and development can have better environmental outcomes than others. Chapter 4 first covers the importance of *where* we build—how safeguarding sensitive areas and focusing development on already-developed land and around transit stations limits environmental impacts. The chapter then covers the importance of *how* we build, looking at different patterns and practices that research indicates are better for the environment, including:

- Compact development.
- Mixed-use development.
- Street connectivity.
- Community design to support walking and biking.
- Development that improves access to destinations and transit.
- Green building.

Finally, Chapter 4 discusses studies that have looked at the combined effects of a range of tools that improve the environmental outcomes of development.

The document concludes by summarizing the effects of how and where we build on efforts to achieve national environmental goals.

Chapter 2. Status of and Trends in Land Use, Buildings, and Travel Behavior

The physical layout and design of our cities and towns have changed dramatically over the past century. In the early 1900s, most urban areas had a compact central business district. Industrial facilities, ports, rail terminals, and other infrastructure hubs anchored major employment centers. Residential areas had small shops and businesses. Suburbs in the first half of the 20th century grew in tandem with extensions of streetcar and railroad lines. They developed in the pattern of the downtown area, on a gridded street network. Each new community typically extended only as far from streetcar lines as people might comfortably walk.

This pattern for cities and towns began to change after World War II, when the United States underwent a period of great economic growth. The expansion of the automobile industry as a key sector of the nation's economy, the launch of the federal interstate highway construction program in 1956, and federal housing policies that encouraged home ownership fundamentally changed the nature of development.[6] The interstate system connected distant locations while also opening up rural areas for development. Metropolitan areas now have multiple clusters of development widely dispersed. Residential, commercial, and industrial uses are separated from each other, often due to local zoning requirements. Many housing subdivisions are located far from stores and services, and if they have sidewalks at all, the sidewalks typically connect only to other homes rather than to places people work, shop, or go to school, making a car essential to daily life. Some business parks are so large that workers need a car to get around inside the park. Shopping malls and strip centers surrounded by large parking lots have replaced many traditional downtown business districts. Street networks are designed to send most traffic to a few large arterial roads. As these changes occurred, people and businesses moved farther from central business districts, while the poor, many of them minorities, were often left behind in communities suffering from disinvestment, where they found it increasingly difficult to access jobs, services, and amenities.[7]

More recently, many city and town centers have revitalized as new people and businesses have moved to these areas for their historic architecture, walkable neighborhoods, and lively street life. As many of the neighborhoods developed before World War II near city centers and along old streetcar lines are being redeveloped, new developments are also starting to adopt designs that appeal to people looking for communities where they can live close to where they work, shop, and take care of other daily needs.

Despite these recent trends, a dispersed pattern of development in the United States dominated during a time in which the population grew significantly. Both population growth and development patterns have contributed to the environmental impacts of our built environment. Development patterns to

---

[6] Vicino 2008
[7] Vicino 2008

accommodate additional population growth in the coming decades will help determine whether these environmental impacts continue to worsen or begin to improve.

This chapter looks at trends in the following areas:

1. Population growth and developed land.
2. Buildings and their water, energy, and materials use.
3. Transportation infrastructure, including roads and parking.
4. Impervious cover.
5. Projected population growth and land conversion.

2.1 Status of and Trends in Population and Developed Land

2.1.1 Population

The population of the United States grew from 76,212,168 in 1900[8] to 311,591,917 in 2011.[9] Most of the population growth occurred in census-defined urban areas, while the rural population remained relatively constant (Exhibit 2-1). In 1900, just 40 percent of the population lived in urban areas; in 2010, more than 80 percent did.

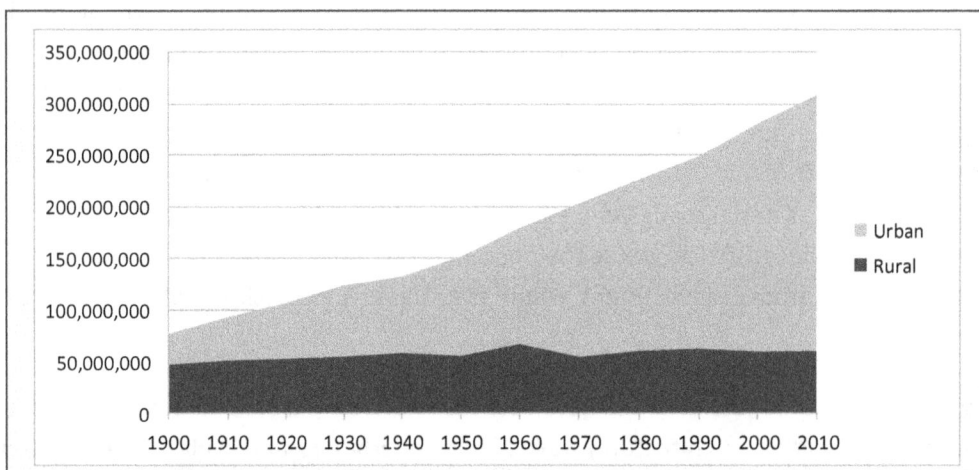

Exhibit 2-1: United States population, 1900-2010. The urban population includes those in urbanized areas of 50,000 or more people and those in urban clusters of at least 2,500 and fewer than 50,000 people. The U.S. Census Bureau does not define suburban population. However, most suburban communities fall under the Census definition of urban population. The rural population includes everyone not in an urban area.
Sources: U.S. Census Bureau *1990 Census of Population and Housing* 1993 and *2010 Census Urban and Rural Classification and Urban Area Criteria* 2010.

---

[8] U.S. Census Bureau, *POP Culture: 1900* n.d.
[9] U.S. Census Bureau, *State & County QuickFacts* n.d.

2.1.2 Metropolitan Area Size

Although the population has grown dramatically in census-defined urban areas, the trend has largely resulted in growth in suburbs rather than in central cities. The proportion of the population residing in the suburbs grew from 23 percent in 1950 to 47 percent in 2010, while the proportion residing in central cities remained relatively constant.[10] A literature review by the National Research Council shows that along with population, employment has shifted out of the central city areas. The combined effect is more spread-out development that makes alternatives to driving (such as transit) less practical.[11] In fact, one of the most dramatic trends in the built environment has been the expansion in the geographic size of metropolitan regions. Virtually every metropolitan region in the United States has expanded substantially in land area since 1950.

Since 1950, the U.S. Census has defined *urbanized areas* to better delineate urban places from rural areas near larger cities. They are determined without regard to political boundaries to better reflect actual settlement patterns. The U.S. Census has modified the definition with each census to ensure that it continues to differentiate in a meaningful way between urban and rural populations. Nevertheless, the U.S. Census has consistently based its definition of urbanized area on a population of at least 50,000 people that are part of a single area primarily based on population density.[12] Exhibit 2-2 shows the population growth and urbanized land area growth between 1950 and 2010 for 39 urbanized areas with populations greater than 1,000,000 in 2010 that were delineated areas in 1950.

For all 39 areas combined, urbanized area increased 2.5 times faster than population growth between 1950 and 2010. This overall trend reflects the fact that urbanized land area in most cities grew at a much faster rate than its population. In urbanized areas that lost population, the contrast between population growth and urbanized area growth is especially remarkable. Eight of the top 15 cities in terms of the ratio of area growth to population growth shown in Exhibit 2-2 had population declines between 1970 and 1980 or between 1980 and 1990: Pittsburgh, Milwaukee, Detroit, St. Louis, Cleveland, New York, Chicago, and Kansas City. In every case, urbanized land area continued to grow during the same period. For example, in 1950, 0.1 acre of land was developed for each resident of the Pittsburgh metropolitan region. At the peak of the region's population in 1970, twice as much land was developed (0.2 acre) for each resident. As population declined, the region continued to expand—by 2010, three times the land was developed for every resident compared to 60 years earlier. Exhibit 2-3 shows how the growth in the metropolitan region's land area continued despite population declines.

However, cities show considerable variation. While the urbanized area of many older, industrial cities such as Pittsburgh, Boston, Milwaukee, Philadelphia, Detroit, St. Louis, and Cleveland grew at a rate more than five times the population growth rate, a minority of cities had population and urbanized area growth trends opposite those of the rest of the country. In these cities, the population growth rate between 1950 and 2010 exceeded the land area growth rate (see Exhibit 2-2). In Los Angeles and San

---

[10] Short 2012

[11] National Research Council of the National Academies, *Driving and the Built Environment* 2009

[12] U.S. Census Bureau, "Notice of final program criteria" 2011

Urbanized Area	Population 1950	Population 2010	Land Area 1950 (mi^2)	Land Area 2010 (mi^2)	Population Growth	Land Area Growth	Ratio of Area Growth to Population Growth
Pittsburgh, PA	1,532,953	1,733,853	253.6	905.17	13%	257%	19.6
Boston, MA–NH–RI	2,233,448	4,181,019	244.8	1873.46	87%	665%	7.6
Milwaukee, WI	829,495	1,376,476	101.7	545.62	66%	436%	6.6
Philadelphia, PA–NJ–DE–MD	2,922,470	5,441,567	311.6	1981.37	86%	536%	6.2
Detroit, MI	2,751,971	3,734,090	422.7	1337.16	36%	216%	6.1
St. Louis, MO–IL	1,400,865	2,150,706	227.8	923.64	54%	305%	5.7
Cleveland, OH	1,383,599	1,780,673	300.1	771.97	29%	157%	5.5
Cincinnati, OH–KY–IN	813,292	1,624,827	146.1	787.74	100%	439%	4.4
Baltimore, MD	1,161,852	2,203,663	151.8	717.04	90%	372%	4.2
New York–Newark, NY–NJ–CT	12,296,117	18,351,295	1,253.4	3450.20	49%	175%	3.6
Indianapolis, IN	502,375	1,487,483	90.6	705.74	196%	679%	3.5
Chicago, IL–IN	4,920,816	8,608,208	707.6	2442.75	75%	245%	3.3
Columbus, OH	437,707	1,368,035	64.5	510.46	213%	691%	3.3
Atlanta, GA	507,887	4,515,419	105.5	2645.35	789%	2407%	3.1
Kansas City, MO–KS	698,350	1,519,417	149.0	677.84	118%	355%	3.0
Jacksonville, FL	242,909	1,065,219	50.8	530.36	339%	944%	2.8
Providence, RI–MA	583,346	1,190,956	142.6	545.05	104%	282%	2.7
Charlotte, NC–SC	140,930	1,249,442	34.5	741.49	787%	2049%	2.6
Washington, DC–VA–MD	1,287,333	4,586,770	178.4	1321.73	256%	641%	2.5
Memphis, TN–MS–AR	406,034	1,060,061	109.6	497.31	161%	354%	2.2
Minneapolis–St. Paul, MN–WI	987,380	2,650,890	231.0	1021.80	168%	342%	2.0
San Antonio, TX	449,521	1,758,210	89.7	597.10	291%	566%	1.9
Seattle, WA	621,509	3,059,393	122.9	1010.31	392%	722%	1.8
Tampa–St. Petersburg, FL	179,335	2,441,770	40.8	956.99	1262%	2246%	1.8
Austin, TX	135,971	1,362,416	34.6	523.03	902%	1412%	1.6
Sacramento, CA	211,777	1,723,634	41.6	470.98	714%	1032%	1.4
Denver–Aurora, CO	498,743	2,374,203	105.2	667.95	376%	535%	1.4
Portland, OR–WA	512,643	1,849,898	113.5	524.38	261%	362%	1.4
Dallas–Fort Worth–Arlington, TX	538,924	5,121,892	142.7	1779.13	850%	1147%	1.3
San Francisco–Oakland, CA	2,022,078	3,281,212	287.3	523.62	62%	82%	1.3
Phoenix–Mesa, AZ	216,038	3,629,114	55.1	1146.57	1580%	1981%	1.3
Orlando, FL	73,163	1,510,516	24.9	597.69	1965%	2300%	1.2
Miami, FL	458,647	5,502,379	116.5	1238.61	1100%	963%	0.9
Houston, TX	700,508	4,944,332	270.1	1660.02	606%	515%	0.8
San Diego, CA	432,974	2,956,746	132.6	732.41	583%	452%	0.8
Salt Lake City–West Valley City, UT	227,368	1,021,243	76.1	277.89	349%	265%	0.8
Riverside–San Bernardino, CA	135,770	1,932,666	60.5	544.97	1323%	801%	0.6
Los Angeles–Long Beach–Anaheim, CA	3,996,946	12,150,996	871.3	1736.02	204%	99%	0.5
San Jose, CA	176,473	1,664,496	60.6	285.98	843%	372%	0.4
Total	**49,629,517**	**130,165,185**	**7,923.7**	**40,206.90**	**162%**	**407%**	**2.5**

Exhibit 2-2: Population growth and land area growth for urbanized areas, 1950-2010. The name of the urbanized area identifies the major place(s) in the urbanized area and the state(s) in which the urbanized area is located.[13]
Sources: U.S. Census Bureau 1961 (Table 22) and *Changes in Urbanized Areas from 2000 to 2010* 2010

Jose, the population growth rate was more than twice the land area growth rate. Cities showing this pattern tend to be newer cities in the Sunbelt with a steadily increasing population.

If population and land area growth for all of the regions in Exhibit 2-2 are examined for 2000 to 2010, rather than 1950 to 2010, to see more recent trends, the overall ratio of urbanized area growth to population growth declines from 2.5 to 1.5. Again, this composite figure obscures noteworthy

---

[13] U.S. Census Bureau, "The Urban and Rural Classifications" 1994

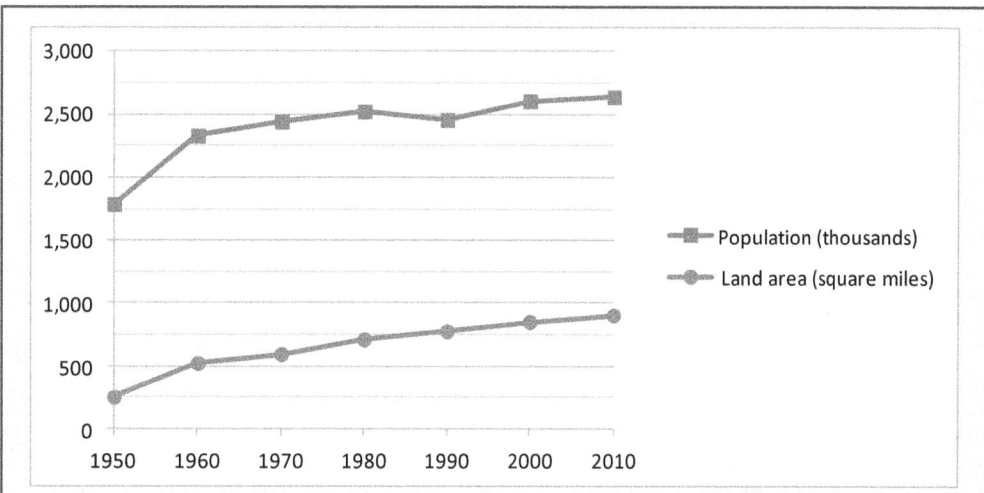

Exhibit 2-3: Pittsburgh metropolitan region land area and population, 1950-2010. Land area includes the amount of land classified as part of the Pittsburgh urbanized area according to the U.S. Census Bureau.
Sources: U.S. Census Bureau 1961 (Table 22), 1979 (Table 1), 1983 (Table 34), 1993 (Table 51), and *Changes in Urbanized Areas from 2000 to 2010* 2010

differences among cities. In this period, the populations of Cleveland, Pittsburgh, and Detroit declined as their urbanized areas continued to expand, and most cities still showed urbanized area growth rates that exceeded population growth rates. However, several regions show the opposite trend: San Diego; San Francisco-Oakland; Seattle; Portland, Oregon; Baltimore; Riverside-San Bernardino, California; Washington, D.C.; Miami; New York-Newark; and Houston. For Baltimore; Washington, D.C.; and New York-Newark, the changing trend is particularly striking, with the number of acres of urbanized area land per resident growing markedly between 1950 and 1980, then leveling off between 1980 and 2010 (Exhibit 2-4).

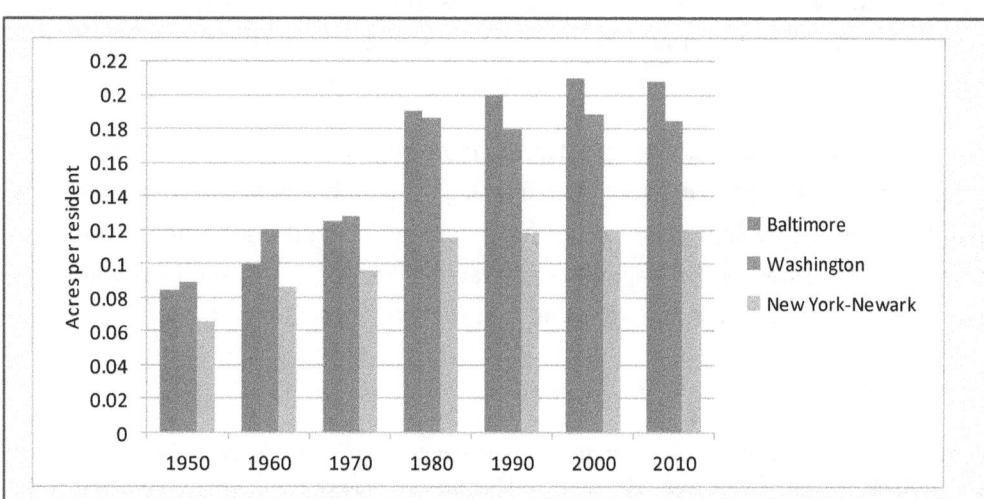

Exhibit 2-4: Urbanized area per resident for the Baltimore, Washington, and New York-Newark urbanized areas, 1950-2010.
Sources: See Exhibit 2-3

Data for all urbanized areas in the United States show that while the urbanized area population and urbanized land area steadily increased between 1950 and 2010, the population per square mile of urbanized areas decreased by more than 50 percent (see Exhibit 2-5). Changes between 1950 and 1980 largely drive this overall trend towards a more dispersed population; population density has remained relatively stable since 1980.

	1950	1960	1970	1980	1990	2000	2010
Urbanized area population	69,252,234	95,848,487	120,726,234	139,170,683	158,258,878	192,323,824	219,922,123
Urbanized area land area (mi^2)	12,805	25,544	36,290	52,017	61,015	72,022	86,776
Population per square mile	5,408	3,752	3,327	2,675	2,594	2,670	2,534

Exhibit 2-5: United States urbanized area changes, 1950-2010. Results are not directly comparable across decades due to changes in the definitions of urbanized areas.
Sources: U.S. Census Bureau 1961 (Table 22), 1979 (Table 1), 1983 (Table 34), 1993 (Table 51), 2004 (Table 6), and *Percent Urban and Rural in 2010 by State* 2010

2.1.3   Developed Land

The U.S. Department of Agriculture publishes a Natural Resources Inventory that tracks developed land using a different definition than the U.S. Census. According to the inventory, as of 1982, close to 71 million acres of non-federal land were developed nationally.[14] By 2007, the number of acres rose to more than 111 million, a 57 percent increase.[15] Over this 25-year period, the population of the United States increased about half as much (30 percent).[16] Of the newly developed land, nearly half (17,083,500 acres) had been forestland in 1982. About one quarter of it (11,117,500 acres) had been cropland, and the remainder had been pastureland, rangeland, or other rural land.[17] Exhibit 2-6 provides an example of how land use changed over four decades in one metropolitan region.

States show considerable variation in the percentage of land that was developed over this time. For example, while North Dakota had just an 8 percent increase in developed land between 1982 and 2007, the increase in Nevada was 145 percent. The other states with the highest percentage increase in developed land over this 25-year period were Georgia (108 percent), North Carolina (107 percent), Florida (99 percent), Arizona (97 percent), and South Carolina (97 percent).[18]

As development occurs, the transition zone between undeveloped and developed land, known as the *wildland-urban interface*, has been expanding, increasing the number of homes surrounded by or adjacent to natural areas. Development that occurs in patches isolated from other developed areas creates new edges on all sides, which fragments habitat (see Section 3.1.3) and can significantly expand the wildland-urban interface. One estimate found that the size of this transition zone grew 52 percent

---

[14] The *National Resources Inventory* category of developed land includes (a) large tracts of urban and built-up land, (b) small tracts of built-up land of less than 10 acres, and (c) land outside of these built-up areas that is in a rural transportation corridor (roads, railroads, and associated rights-of-way).
[15] U.S. Department of Agriculture 2009
[16] Calculated based on data from the U.S. Census Bureau (U.S. Census Bureau, *Population and Housing Unit Estimates* n.d.).
[17] U.S. Department of Agriculture 2009
[18] U.S. Department of Agriculture 2009

between 1970 and 2000.[19] In the western United States, nearly 90 percent of this interface occurs in areas with a high risk of forest fires, making the growth of housing in and near this zone of particular concern.[20]

Data on the extent of land developed does not fully capture development's impacts because some areas are more sensitive to its effects than others are. One study found that housing in and near wilderness, national parks, and national forests has increased since 1940. The number of homes located within 1 kilometer (0.62 miles) of these protected areas increased nearly fourfold, from slightly less than half a million in 1940 to slightly less than 2 million in 2000. New residential construction *within* national forests rose from 335,000 units in 1940 to nearly 1.3 million units 60 years later, adding the equivalent of three homes per square mile of national forest.[21]

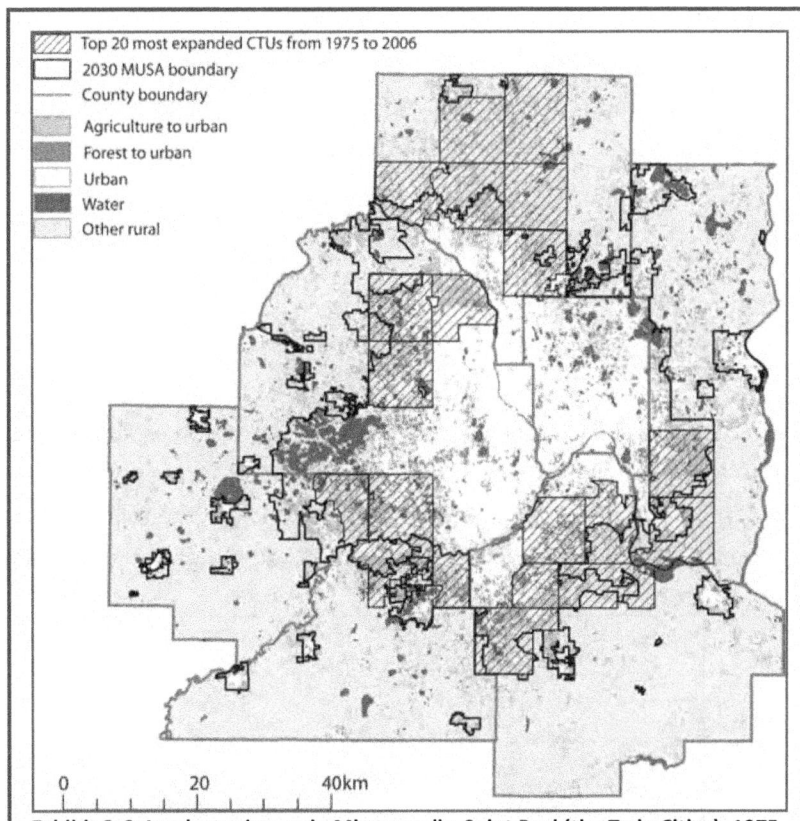

Exhibit 2-6: Land use change in Minneapolis–Saint Paul (the Twin Cities), 1975-2006. A study of land use change in the Twin Cities metropolitan urban service area (MUSA) between 1975 and 2006 showed that urban area increased by 313,100 acres (about 83 percent), mostly in areas previously covered by forest, cropland, or wetlands. Most new growth occurred along major highways and roads at the periphery of the urban area or in new growth centers disconnected from the core. The seven counties that comprise the metropolitan area had a population growth rate of 45 percent during this same period, which corresponds to an annual growth rate of 1.5 percent, making this the eighth fastest growing area in the United States during that period. Shaded areas in a ring around the region show the top 20 most expanded cities, townships, and unorganized territories (CTUs) from 1976 to 2006.
Image source: Yuan 2010. Reprinted by permission of the publisher (Taylor & Francis Ltd, *www.tandf.co.uk/journals*).

Coastal ecosystems are also particularly sensitive to the effects of development. Although coastal watershed counties constitute 20 percent of total U.S. land area (excluding Alaska), in 2010 they contained 52 percent of the U.S. population and had an average density of 319 people per square mile compared with 61 people per square mile in inland counties.[22] A vulnerability index created by the U.S.

[19] Theobald and Romme 2007
[20] Massada, et al. 2009
[21] Radeloff, et al. 2010
[22] National Oceanic and Atmospheric Administration 2013

Geological Survey[23] identifies 6,734 miles (30 percent) of coastal shoreline in the United States as very highly vulnerable to sea level rise and an additional 4,514 miles (20 percent) as highly vulnerable.[24] Between 1970 and 2010, the population in the coastal flood plain increased 51 percent. As of 2010, 16.4 million people lived in the coastal flood plain, an area at greater risk of inundation from storm surges and long-term sea level rise.[25]

2.2   Status of and Trends in Buildings

2.2.1   Housing Units

The housing industry is highly susceptible to market forces, and year-to-year fluctuations tend to track overall economic health. After peaking in 2006 and surpassing the previous high set during the mid-1980s, the number of new homes fell precipitously in the following years (Exhibit 2-7). About 700,000 new housing units were built in 2010, compared to more than 2 million units four years earlier.[26]

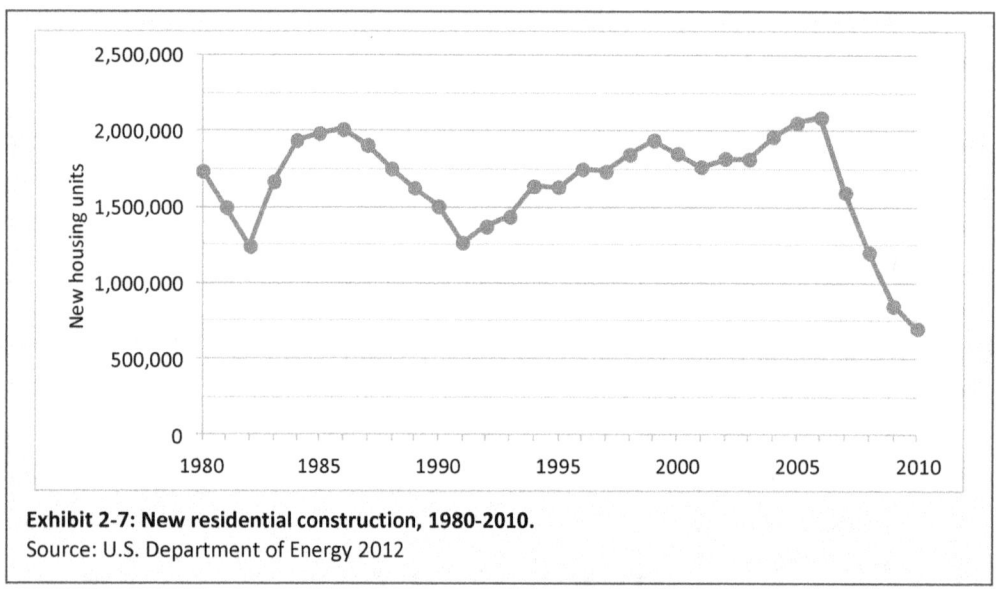

Exhibit 2-7: New residential construction, 1980-2010.
Source: U.S. Department of Energy 2012

In spite of fluctuations in the number of new homes produced annually, the total number of homes in the United States has grown steadily since 1940 (Exhibit 2-8). Between 1940 and 2010, the national housing inventory grew more than 250 percent, adding nearly 100 million homes, while population grew by 134 percent or 176,580,969 people (Exhibit 2-1).

---

[23] U.S. Geological Survey, *National Assessment of Coastal Vulnerability to Sea-Level Rise* n.d.
[24] National Oceanic And Atmospheric Administration, *Climate: Vulnerability of our Nation's Coasts to Sea Level Rise* n.d.
[25] National Oceanic and Atmospheric Administration, *Climate: U.S. Population in the Coastal Floodplain* n.d.
[26] U.S. Department of Energy 2012

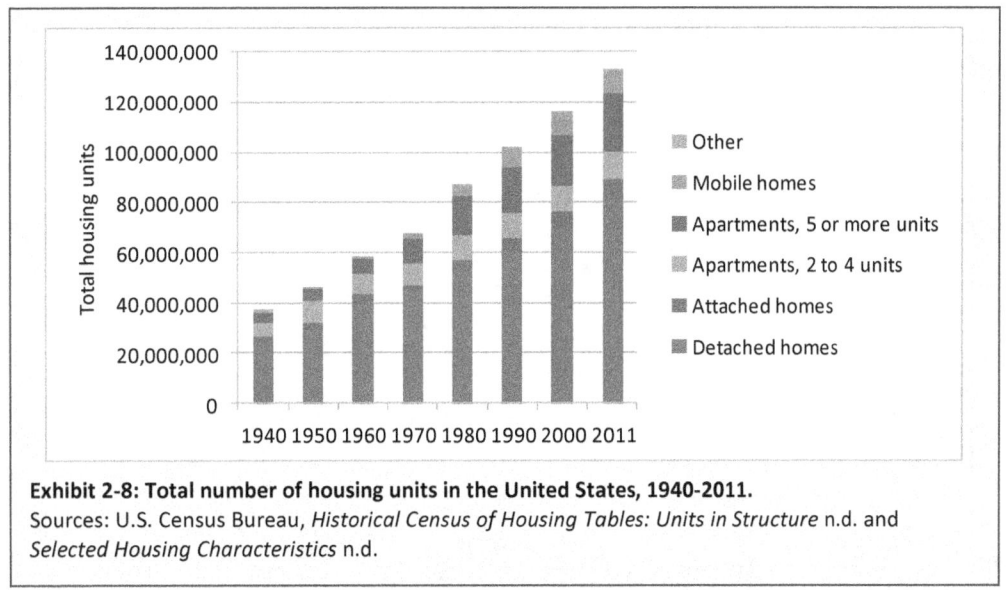

Exhibit 2-8: Total number of housing units in the United States, 1940-2011.
Sources: U.S. Census Bureau, *Historical Census of Housing Tables: Units in Structure* n.d. and *Selected Housing Characteristics* n.d.

As the number of housing units increased, so did the average size of houses. Single-family homes built between 2000 and 2005 are 29 percent larger than homes built in the 1980s and 38 percent larger than homes built before 1950.[27] The national average for single-family residences went from 1,660 square feet in 1973 to more than 2,500 square feet in 2007 before dropping slightly in subsequent years (Exhibit 2-9).

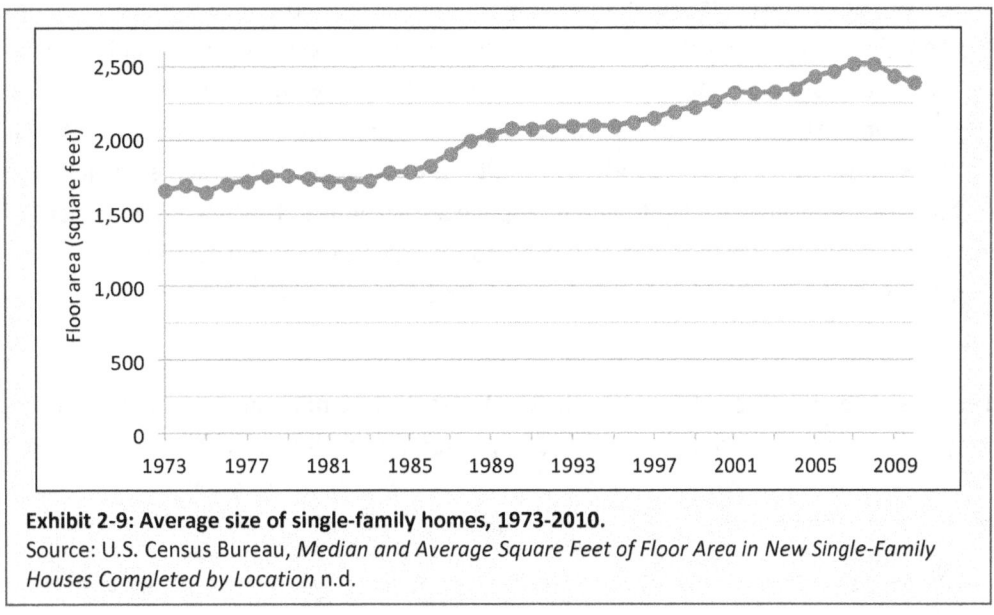

Exhibit 2-9: Average size of single-family homes, 1973-2010.
Source: U.S. Census Bureau, *Median and Average Square Feet of Floor Area in New Single-Family Houses Completed by Location* n.d.

The trend toward larger homes has occurred while average household size has decreased. The average U.S. household was 2.59 persons in 2010, compared with 3.01 in 1973 and 3.56 in 1947 (Exhibit 2-10).[28] Shifts in the composition of households drove this trend. In 1940, only 8 percent of households

---

[27] U.S. Department of Energy 2012
[28] U.S. Census Bureau, *Current Population Survey* 2011

contained only one person, and 51 percent of households had children under 18. By 2011, one-person households had increased to 28 percent, and households with children had dropped to 29 percent. [29]

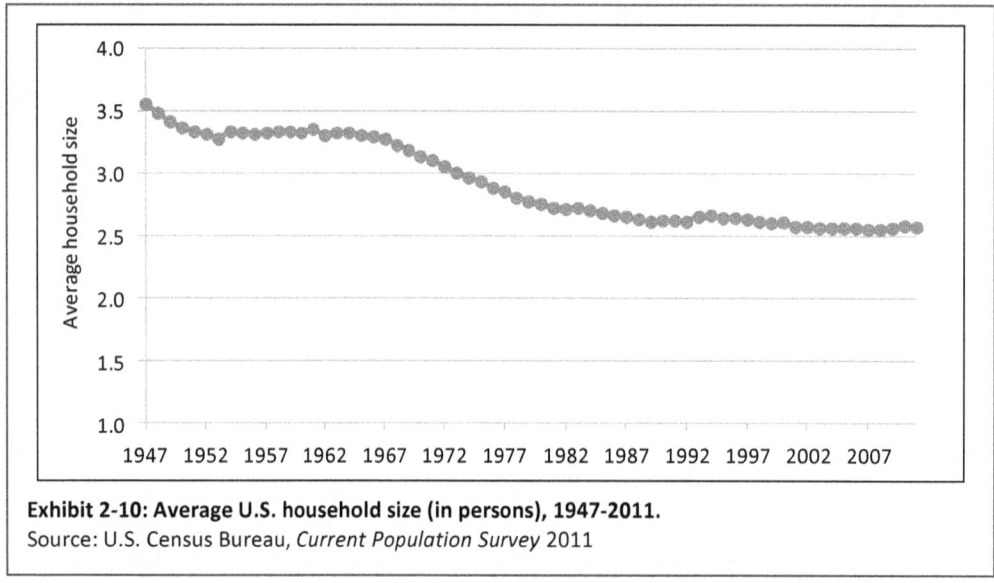

Exhibit 2-10: Average U.S. household size (in persons), 1947-2011.
Source: U.S. Census Bureau, *Current Population Survey* 2011

2.2.2   Building Energy Use

In 2011, the U.S. building sector[30] accounted for 41 percent of domestic energy use and 7 percent of global energy use.[31] Energy use has grown relatively steadily for both the residential and commercial sectors since the 1950s, although the residential sector has more year-to-year fluctuations (Exhibit 2-11), due to weather, consumer income, and overall economic activity. The building sector's energy use quadrupled between 1950 and 2010 while the population roughly doubled (as shown in Exhibit 2-1) and the housing stock more than tripled (as shown in Exhibit 2-8). In 2010, purchased electricity for space heating and cooling, water heating, lighting, and appliances accounted for 71 and 78 percent of total building energy consumption for the residential and commercial sectors, respectively. Direct consumption of natural gas and petroleum for heating and cooking accounts for the remaining energy use.[32]

Buildings rely heavily on electricity for power. In 2010, just over 10 percent of the power generated by the electricity sector came from renewable sources, predominantly hydroelectric.[33] After purchased

---

[29] U.S. Census Bureau, *1940 Census of Population and Housing—Families* 1943, *Historical Census of Housing Tables: Living Alone* n.d., and *American FactFinder* n.d.

[30] In this document, the building sector refers to residential and commercial buildings only. The residential building sector includes single- and multi-family residences. The commercial building sector includes offices, stores, restaurants, warehouses, other buildings used for commercial purposes, and government buildings.

[31] U.S. Department of Energy 2012

[32] EPA, *Inventory of U.S. Greenhouse Gas Emissions and Sinks* 2012

[33] U.S. Energy Information Administration 2011. The percentage of power generated by the electricity sector that comes from renewable sources has varied between 7 and 14 percent since 1976. The share was 31 percent in 1949 and fell for the next 28 years.

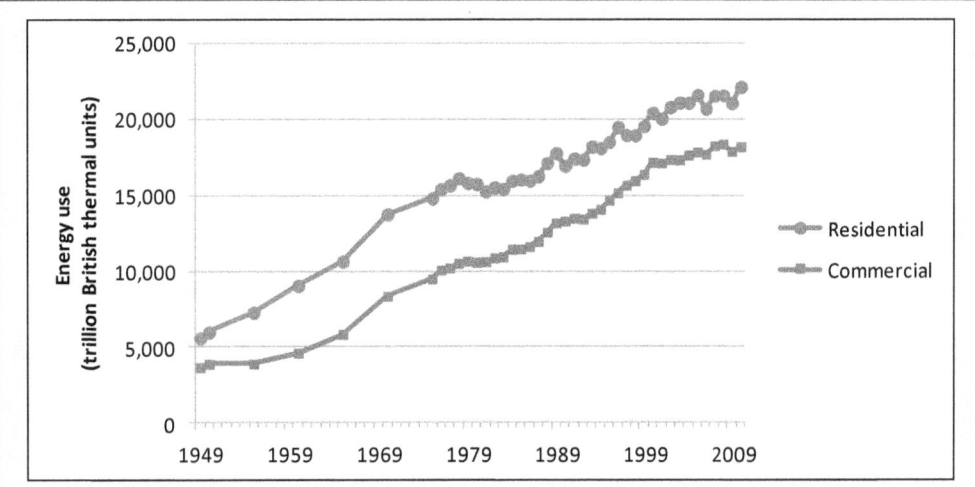

Exhibit 2-11: Energy use in British thermal units (Btu) by building sector, 1949-2010. Energy use includes primary energy consumption, electricity retail sales, and electrical system energy losses.
Source: U.S. Energy Information Administration 2011, Table 2.1a

electricity, burning fossil fuels (i.e., natural gas, petroleum, and coal) is the next most common energy source for buildings in both the residential (Exhibit 2-12) and commercial sectors (Exhibit 2-13). Between 1949 and 2010, fossil fuels accounted for between 76 and 95 percent of residential direct energy use, with no clear trend over time. For the commercial building sector, essentially all direct energy prior to 1988 came from fossil fuels, and use of renewable resources increased only slightly in later years. In 2010, just 3 percent of direct energy in the commercial building sector came from renewable resources, predominantly wood.[34]

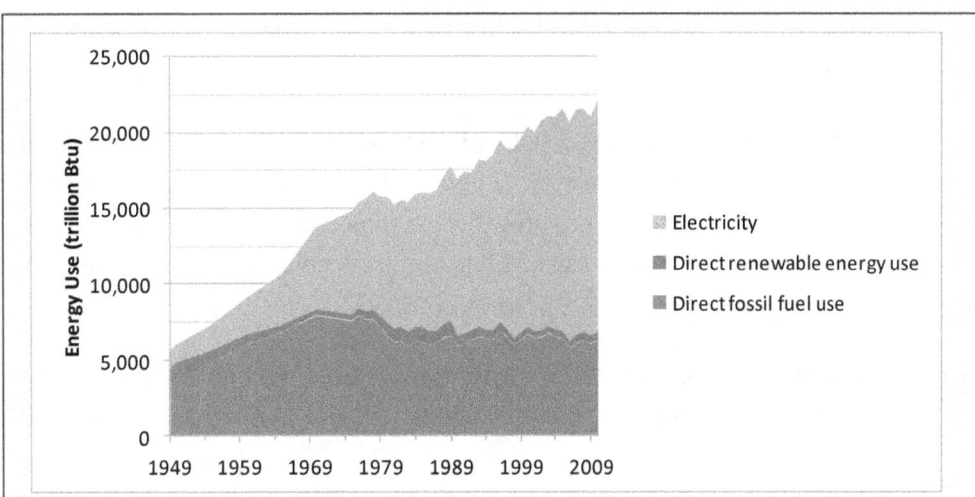

Exhibit 2-12: Energy use from fossil fuels for residential buildings, 1949-2010. Renewable energy sources include geothermal, solar, photovoltaic, and wood. Electricity includes electricity retail sales and electrical system energy losses.
Source: U.S. Energy Information Administration 2011, Table 2.1b

---

[34] U.S. Energy Information Administration 2011

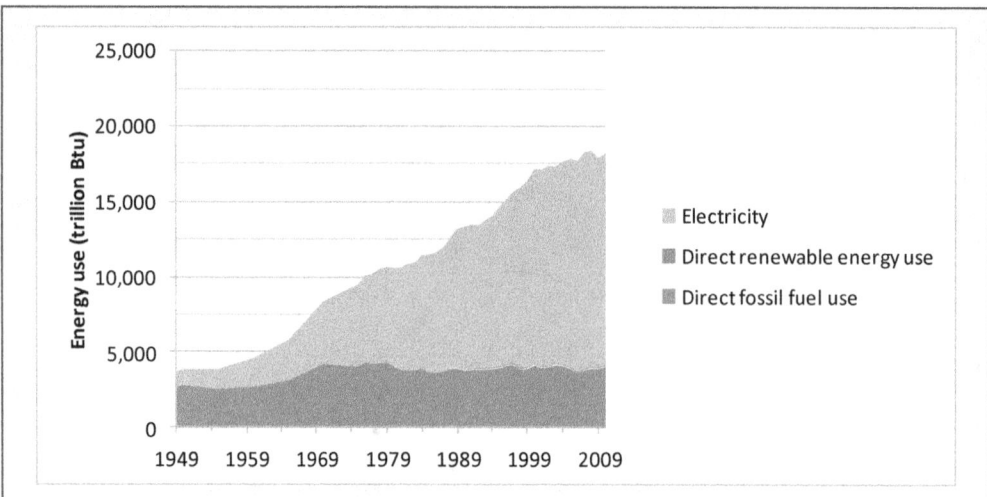

Exhibit 2-13: Energy use from fossil fuels for commercial buildings, 1949-2010. Renewable energy sources include geothermal, solar, photovoltaic, wood, hydroelectric, and wind. Electricity includes electricity retail sales and electrical system energy losses.
Source: U.S. Energy Information Administration 2011, Table 2.1c

The generation of energy used for buildings also requires water. Much of it is needed for cooling at power plants.[35] In 2005, fresh water withdrawals for electricity generation totaled 510 million gallons from ground sources and 142 billion gallons from surface sources.[36] Total water withdrawals including saltwater were 201 billion gallons, which was the amount needed to generate 3.19 trillion kilowatt hours of electricity.[37] Thus, every kilowatt hour of electricity (enough energy to power a 100-watt light bulb for 10 hours) requires 16 gallons of water, and 71 percent of that water comes from ground water, lakes, and rivers.

2.2.3   Building Water Use

Buildings get water from the public supply and self-supplied sources such as surface water, wells, and rainwater cisterns.[38] Water consumption grew by 27 percent in the building sector between 1985 and 2005[39] (Exhibit 2-14), even as overall water use grew by less than 3 percent.[40] In 2005, water consumption in the residential and commercial building sectors accounted for 9.7 percent of all water consumption in the United States, up from 7.8 percent in 1985.[41]

Earlier data are available tracking combined water withdrawals from the public supply (sourced from both ground water and surface water) for residential, industrial, and commercial use, as well as public

---

[35] Kenny, et al. 2009
[36] Kenny, et al. 2009
[37] Kenny, et al. 2009
[38] Kenny, et al. 2009
[39] The U.S. Geological Survey does not expect to release source data for 2010 until 2014 (U.S. Geological Survey, *Water Use in the United States* n.d.).
[40] U.S. Department of Energy 2012
[41] U.S. Department of Energy 2012

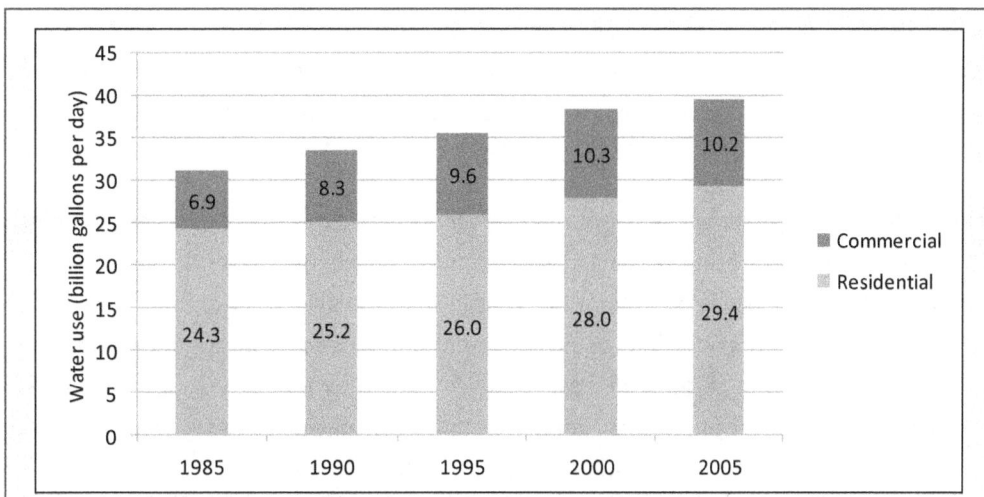

Exhibit 2-14: Water use in the building sector, 1985-2005. For the years 2000 and 2005, the split between commercial and residential use is based on extrapolation from 1995 data.
Source: U.S. Department of Energy 2012

services and system losses. Per capita withdrawals between 1950 and 1980 increased more than 60 percent but then largely leveled off between 1985 and 2005 (Exhibit 2-15). In 2005, on average, 58 percent of the total withdrawals from the public supply were for indoor and outdoor residential use, although results varied by state, from 39 percent in Pennsylvania to 79 percent in Maryland.[42]

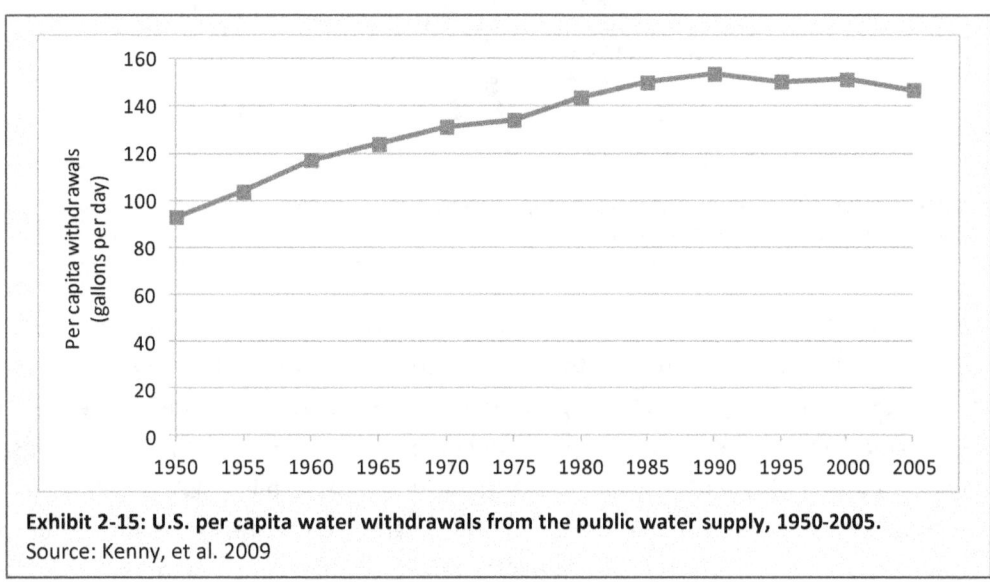

Exhibit 2-15: U.S. per capita water withdrawals from the public water supply, 1950-2005.
Source: Kenny, et al. 2009

Water use in commercial buildings varies by season and depends on a host of factors, including the building's size, function, and location. In residential buildings, water use averages about 100 gallons per person per day.[43] However, per capita residential water use varies considerably by state, driven in large measure by greater outdoor water use in arid regions. In 2005, the lowest per capita use was in Maine

---

[42] Kenny, et al. 2009
[43] U.S. Department of Energy 2012

(54 gallons per day), Pennsylvania (57 gallons per day), Wisconsin (57 gallons per day), Delaware (61 gallons per day), and Vermont (64 gallons per day). The highest per capita use was nearly triple these rates in Nevada (190 gallons per day), Idaho (187 gallons per day), Utah (186 gallons per day), Hawaii (165 gallons per day), and Wyoming (152 gallons per day).[44]

Buildings and building occupants use water for various indoor and outdoor activities including drinking; bathing; cleaning; cooking; watering lawns and landscaping; supplying fountains and other water features; and running cooling towers for heating, ventilating, and air conditioning. An analysis of 1,188 single-family homes showed that outdoor uses (e.g., lawn and landscaping irrigation and swimming pools) accounted for about one-third of all water used, followed by toilets, washing machines, faucets, and showers (Exhibit 2-16),[45] although this likely varies considerably depending on climate. Exhibit 2-17 shows how several types of commercial and institutional facilities use water.

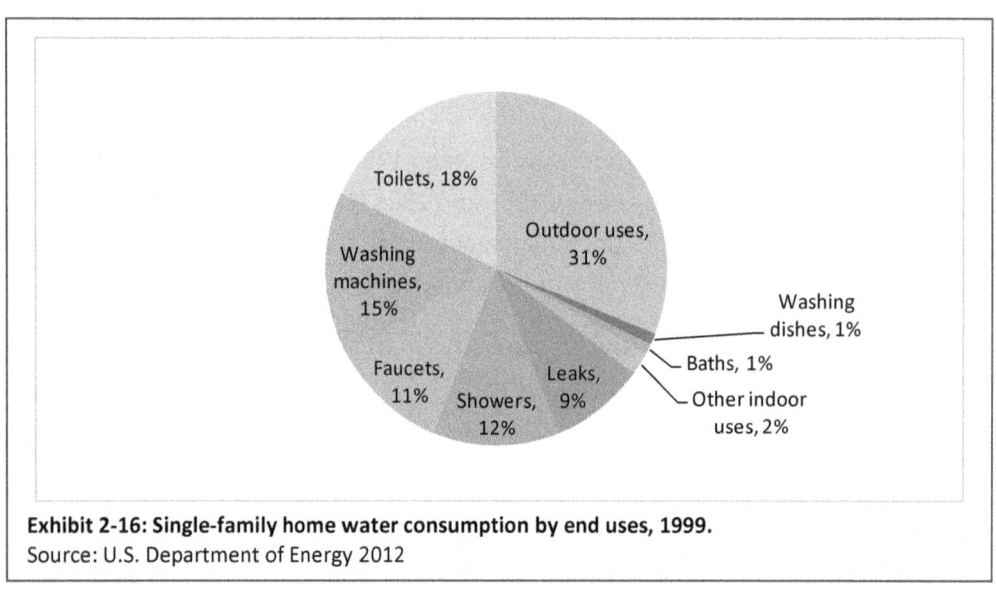

Exhibit 2-16: Single-family home water consumption by end uses, 1999.
Source: U.S. Department of Energy 2012

Water use affects more than just the amount of water withdrawn from underground aquifers, surface water bodies, and other sources. Water treatment and distribution are highly energy-intensive, as are many water uses that involve heating, chilling, softening, or pressurizing it for various needs.[46] Uneven data availability across states and methodological difficulties make analysis of the energy consumption associated with water use difficult.[47] One analysis estimated it took 39 billion kilowatt hours of electricity to supply water to the building sector in 2005, a value that represented about 1 percent of total electricity produced by all power plants in the United States that year.[48] Another study estimated that the residential sector consumes 2,747 trillion British thermal units (Btu) of energy annually for water services, primarily for heating water, while the commercial sector consumes 1,657 trillion Btu of

---

[44] Kenny, et al. 2009
[45] U.S. Department of Energy 2012
[46] Sanders and Webber 2012
[47] Sanders and Webber 2012
[48] U.S. Department of Energy 2012

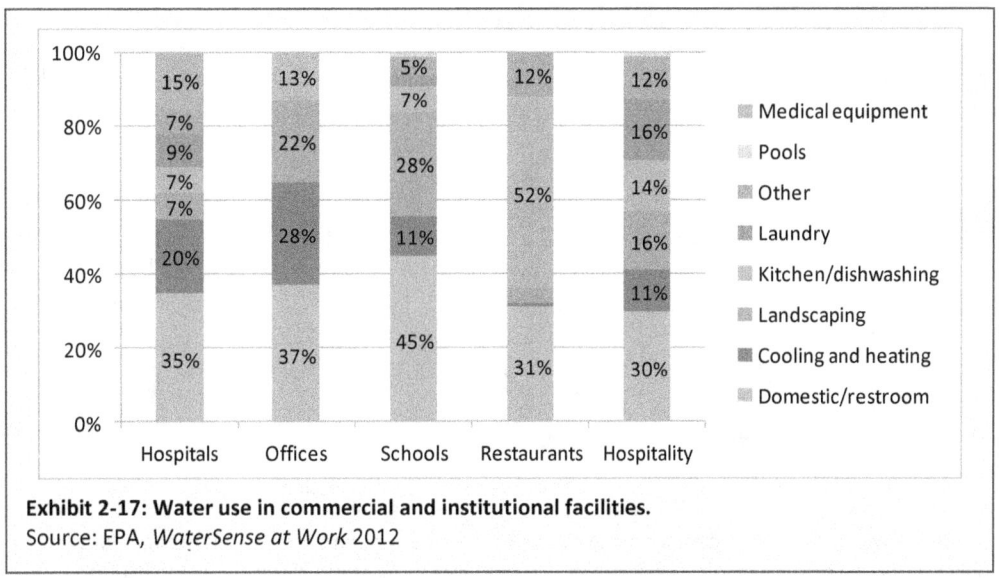

Exhibit 2-17: Water use in commercial and institutional facilities.
Source: EPA, *WaterSense at Work* 2012

energy annually for water services, primarily at public water and wastewater treatment facilities and for heating water. Based on these figures, combined water use in the residential and commercial sectors accounted for roughly 4.5 percent of national energy consumption in 2010.[49]

2.2.4   Building Construction Waste Production

Buildings use a lot of construction materials that result in an equivalent amount of demolition debris during renovations and at the end of their lifecycle. Much of this material is shipped to landfills specifically designated for construction and demolition waste, while in some areas it is discarded with municipal solid waste in landfills, used as fill in quarries or pits, or incinerated.[50] Cement concrete alone accounts for about half of the 65 million tons of demolition debris produced each year.[51] EPA estimated that in 2003 the building sector produced an estimated 170 million tons of construction, renovation, and demolition waste.[52] Another estimate of construction and demolition debris from the building sector—based on the total amount of construction material used over time, its typical service life, and the amount of material discarded during construction—was 209 million tons in 2002.[53] Construction and demolition waste includes wood, drywall, asphalt, concrete, cardboard, glass, masonry, roofing material, plastic, and metal. Commercial renovation was the leading source of construction and demolition waste in 2003, followed by demolition waste from commercial and residential buildings (Exhibit 2-18). Although information is not available on trends in the amount of construction and demolition waste produced over time, as the size of single-family homes has increased (Exhibit 2-9), the amount of materials used for home construction and the amount of waste that will be generated at the end of the homes' lifecycle have likely also increased.

---

[49] Sanders and Webber 2012
[50] EPA, *Estimating 2003 Building-Related Construction and Demolition Materials Amounts* 2009
[51] Horvath 2004
[52] EPA, *Estimating 2003 Building-Related Construction and Demolition Materials Amounts* 2009
[53] Cochran and Townsend 2010

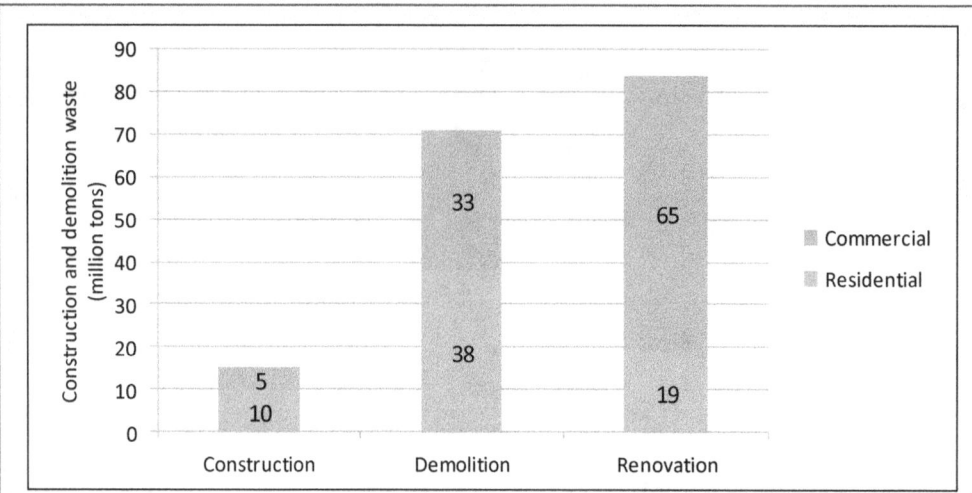

Exhibit 2-18: Amount of construction and demolition waste generated from building activities, 2003. Although renovation of a building produces less waste than demolition and construction of a new building, considerably more renovation activity occurs, making it a larger contributor to overall amounts of construction and demolition waste.
Source: EPA, *Estimating 2003 Building-Related Construction and Demolition Materials Amounts* 2009, Table 2-7

While in aggregate, new construction generally produces less construction and demolition waste than demolition and renovation, the construction of a new single-family detached residence can generate up to 7 tons of waste and 15 to 70 pounds of potentially hazardous substances such as paint, caulk, roofing cement, aerosols, solvents, adhesives, oils, and greases. One study estimated that in the U.S. residential building sector, construction contributes more hazardous waste and toxic air emissions than building use or demolition based on estimates of the amount of construction and demolition activity occurring in a single year (1997).[54]

2.3 Status of and Trends in Infrastructure

The increase in developed land and buildings in the United States has occurred along with an increase in transportation and other infrastructure needed to serve an increasingly dispersed population.

2.3.1 Roads

Roads alone represent a considerable portion of the built environment. In 2010, the United States had more than 4 million miles of roads owned and maintained by a public authority and open to public travel (Exhibit 2-19). The amount of annual road construction was greatest during the 1910s and then

---

[54] Ochoa, Hendrickson, and Matthews 2002. Construction included acquisition of raw materials, manufacturing, transportation to the construction site, and actual construction. The study compared this phase with usage, which included remodeling, improvements, and occupant energy usage, and demolition, which included the demolition process and debris transportation.

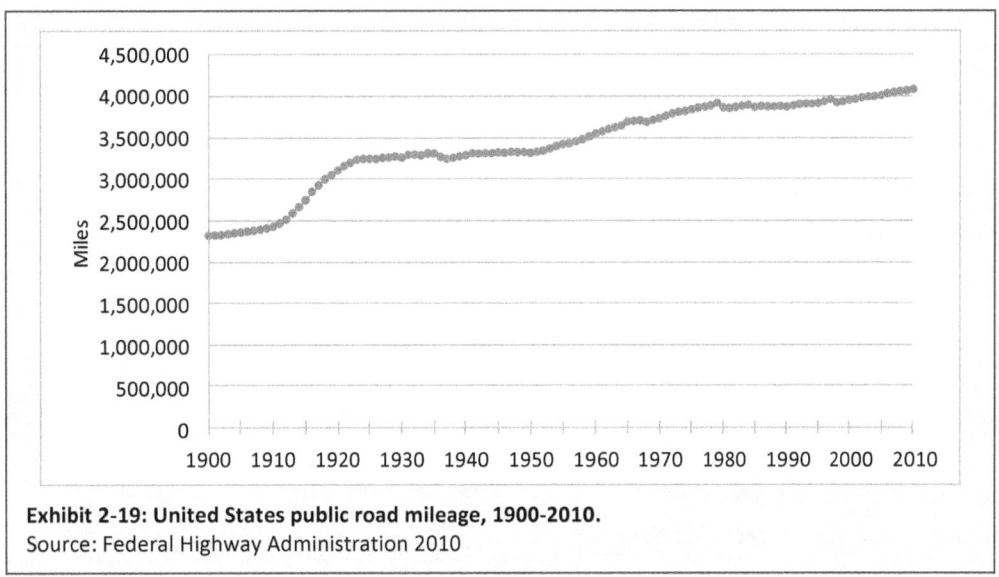

Exhibit 2-19: United States public road mileage, 1900-2010.
Source: Federal Highway Administration 2010

leveled off until another period of expansion in the 1950s and 1960s. Although the rate of new road construction has slowed since then, 114,576 lane miles of public roads were built between 2000 and 2005, and another 277,385 lane miles were added in the following five years.[55]

The total amount of land covered by roads does not capture their full impact because their environmental effects extend beyond the pavement edge (see Section 3.1.4). As more lane miles are built, the amount of land they affect increases. In 2001, 20 percent of all land in the contiguous United States was within 417 feet of a road, and 50 percent was within a quarter-mile. Only about 18 percent of all land was more than 0.62 mile from a road, and about 3 percent was more than 3 miles.[56]

2.3.2 Parking

Parking structures and parking lots are also significant components of our transportation infrastructure in terms of the amount of land they cover. In 2010, more than 242 million cars, buses, and trucks were registered in the United States,[57] and each requires a place to park at home and other destinations. Estimates of the total parking inventory in the United States vary widely due to differences in methodology. A 2010 study developed five scenarios based on existing literature to estimate the total number of parking spaces and the land area they consumed. The middle scenario was based on the number of known for-pay parking spaces, estimates of the amount of parking provided per square foot of commercial space, estimates of on-street parking based on road design specifications, and assumptions that each car had a home space and work space. This scenario estimated 820 million parking spaces covering 8,500 square miles.[58] If combined with the amount of space devoted to roads, an estimated 23,900 square miles of the United States is paved for driving and parking, an area nearly

---

[55] Federal Highway Administration 2010
[56] Riitters and Wickham 2003
[57] Federal Highway Administration 2010
[58] Chester, Horvath, and Madanat 2010

the size of West Virginia.[59] More precise estimates have been made in smaller regions using satellite images. For example, a study of Tippecanoe County, Indiana, found that parking alone covers 2.2 square miles, 0.4 percent of the total land and 6.6 percent of the developed area in the county.[60]

Exhibit 2-20: Parking Lot. Parking lots, many of which regularly sit empty, cover significant amounts of land.
Photo source: EPA

2.3.3   Water, Wastewater, Utilities, and Other Infrastructure

Reliable national data on the extent of centralized water and wastewater, utilities, and other infrastructure and how they have changed over time are unavailable. One estimate, based on a survey of utilities and extrapolated based on the population they served, estimated that the United States has 995,644 miles of pipes in its water distribution network.[61] Another analysis estimated the United States has 800,000 miles of drinking water pipes and between 600,000 and 800,000 miles of wastewater pipes.[62] As noted in Section 2.1.2, as the population has grown, population density has decreased in nearly all metropolitan regions. With this change, the amount of centralized infrastructure needed per person has likely also increased as distances between houses have grown.

In addition to materials needed for construction, repair, and replacement of infrastructure, centralized water and wastewater infrastructure itself has major environmental effects. The energy needed to move water and wastewater increases as centralized systems cover more land. A larger system also creates more opportunity for treated drinking water and untreated wastewater to be lost through leaks because of both the length of the pipes and the additional pressure needed to push water through longer pipes.

Many households rely on decentralized water and wastewater infrastructure, and this pattern continues for newly built homes. In 2011, 11 percent of all households and 11 percent of households in a home built in the last four years used a well as their primary source of water. For wastewater infrastructure, 19 percent of all households and 21 percent of households in a home built in the last four years relied on a septic tank, cesspool, or chemical toilet.[63] These types of decentralized wastewater systems can release nitrogen and phosphorus to both ground and surface water. Although nitrogen and phosphorus are important nutrients, in excess they can contribute to an overgrowth of aquatic plant life that depletes oxygen other aquatic species need to survive. Conventional septic systems remove less than

[59] Chester, Horvath, and Madanat 2010
[60] Davis, et al. 2010
[61] EPA, *Distribution System Inventory, Integrity and Water Quality* 2007
[62] U.S. Government Accountability Office 2004
[63] U.S. Census Bureau, *American FactFinder* n.d., Table C-04-AO

half of the nitrogen from wastewater, although advanced treatment systems can remove much more.[64] Although septic systems release treated wastewater to soils, which can better retain phosphorus, septic systems can still lead to phosphorus discharges to ground and surface water depending on soil characteristics, septic system placement, and use of phosphorous-containing household products.[65] In addition, improperly maintained systems can discharge untreated or inadequately treated wastewater.[66]

2.4   Status of and Trends in Impervious Cover

Land development—the buildings and roads we construct for housing, commercial activity, industry, and transportation—creates impervious surfaces that water cannot penetrate. Under natural conditions, only about 10 percent of the annual rainfall becomes surface runoff; the rest either soaks into the soil or is taken up by vegetation and transpired.[67] As development occurs, impervious surfaces—including roofs, roads, driveways, sidewalks, and patios—replace vegetation, and more and more rainwater travels as runoff from the area where it fell. The environmental effects of this change are varied and can be significant even if only a small proportion of a watershed is developed (see Section 3.2). The proportion of a watershed that is covered with impervious surface can serve as a good indicator of the degree to which development affects water ecosystems and water resources.[68]

A national estimate of impervious cover in the contiguous United States (total area 3,035,033 square miles) found that the total impervious surface area in 2006 was 40,006 square miles, an area slightly smaller than the state of Kentucky, and a 4 percent increase since 2001.[69] The states with the largest percentage increases were Arizona (8.9 percent), Georgia (8.4 percent), and South Carolina (7.9 percent). The states with the largest absolute increases in impervious surface area during this five-year period were Texas (212 square miles, a 6.0 percent increase), California (122 square miles, a 3.7 percent increase), and Florida (107 square miles, a 5.7 percent increase). In most places, most new impervious cover formed a ring around central cities.

An earlier analysis estimated the amount of impervious surface in the contiguous United States to be 32,317 square miles in 2000.[70] The study used a housing density model to estimate the distribution of this impervious cover among watersheds. It found that impervious cover was at levels likely to affect water quality in almost 25 percent of watersheds by area (Exhibit 2-21).

Although many large watersheds still contain relatively small percentages of impervious cover, overall impervious cover is not necessarily a good indicator of whether there are local effects. Impervious cover

---

[64] Carey, et al. 2013

[65] Carey, et al. 2013

[66] Carey, et al. 2013

[67] Federal Interagency Stream Restoration Working Group 1998

[68] Sutton, et al. 2009

[69] Xian, et al. 2011. This estimate is based on the U.S. Geological Survey's National Land Cover Database.

[70] Theobald, Goetz, et al. 2009. This estimate is comparable to others and supported by validation data sets, although the methodology used is known to under-represent impervious cover in areas with commercial and industrial land uses. The estimate for 2000 is 16 percent less than Xian, et al. estimated for 2001.

Watershed classification	Percent impervious cover	Number of watersheds	Percent by area
Unstressed	0-0.9	45,098	75.5
Lightly stressed	1-4.9	13,020	20.3
Stressed	5-9.9	1,305	1.8
Impacted	10-24.9	773	0.9
Damaged	>25	197	0.2

Exhibit 2-21: Number and percentage by area of watersheds affected by impervious cover levels, 2000. For this analysis, watershed boundaries were determined by 10-digit hydrologic unit codes, a classification system that divides the country into nested hydrologic units. Source: Theobald, Goetz, et al. 2009

is often concentrated in particular areas, and smaller watersheds contained within larger watersheds can have much higher relative levels. For example, most cities have levels of impervious cover that are known to stress or even seriously degrade water quality.[71] Estimates of impervious cover for 18 cities throughout the United States ranged from 61 percent in New York City to 18 percent in Nashville.[72] The percentage of impervious cover across all of these cities grew in the mid to late 2000s on average by 0.31 percent per year, or 23.7 square feet per person per year, at the same time that these cities lost tree canopy cover. For example, Houston's impervious cover grew on average 0.26 percent while it lost

0.60 percent of its tree cover every year between 2004 and 2009. Tacoma, Washington's impervious cover grew on average 0.89 percent while it lost 0.36 percent of its tree cover every year between 2001 and 2005. Syracuse, New York, is the only one of the 18 cities that both increased tree cover, which grew on average by 0.17 percent per year from 2003 to 2009, and reduced impervious cover, which declined by 0.09 percent per year. The increase in tree cover in Syracuse was due predominantly to growth of invasive European buckthorn, suggesting that natural regeneration and limited development activity is responsible for the change. Exhibit 2-22 shows an example of how the amount of impervious cover changed in two Washington counties over 34 years.

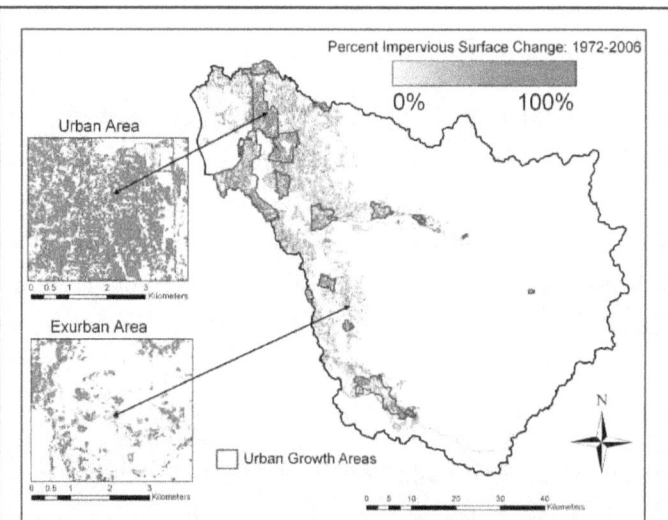

Exhibit 2-22: Impervious cover change in Snohomish and King counties, Washington. Satellite imagery for a portion of Snohomish and King counties, Washington, was analyzed to estimate the increase in impervious cover between 1972 and 2006. Over this 34-year period, the area of impervious surface increased by 255 percent (8,117 to 28,793 acres), while population increased by 79 percent. About three-quarters of the new impervious area was added in relatively low-density locations that were outside of the counties' designated urban growth areas (outlined in blue). Image source: Powell, et al. 2008. Reprinted by permission of the publisher (Elsevier).

---

[71] Theobald, Goetz, et al. 2009

[72] Nowak and Greenfield 2012. Cities were selected based on available data and existing projects and for geographic representation across the country. City boundaries were defined by census-incorporated place or census-designated place boundaries. The years analyzed for each city vary slightly based on the dates of available aerial images.

2.5   Status of and Trends in Travel Behavior

Numerous factors have contributed to an increase in vehicle travel since the 1950s. For example, the size of the U.S. labor force affects how many people commute to work in cars. The U.S. labor force was roughly 59 percent of the working-age population in 1948.[73] It remained relatively constant until the late 1960s. From then until 2000, it grew to a peak of 67 percent, then declined slightly and rose again to 66 percent in 2006. A substantial portion of the increase in labor force participation until 2000 was due to women, particularly married women, joining the workforce. An increasing Hispanic population also explains some of the increase, as Hispanic men tend to have the highest workforce participation rates.[74]

A combination of rising incomes and falling fuel prices in the latter half of the twentieth century also affected vehicle travel. As household incomes increased and fuel prices fell, families could afford one or more cars. In 1969, 79 percent of U.S. households owned one or more vehicles. By 1995, that figure had grown to 92 percent. However, between 2001 and 2009 the number of households with no vehicle grew by nearly 1 million, rising from 8.1 to 8.7 percent of all households (Exhibit 2-23).

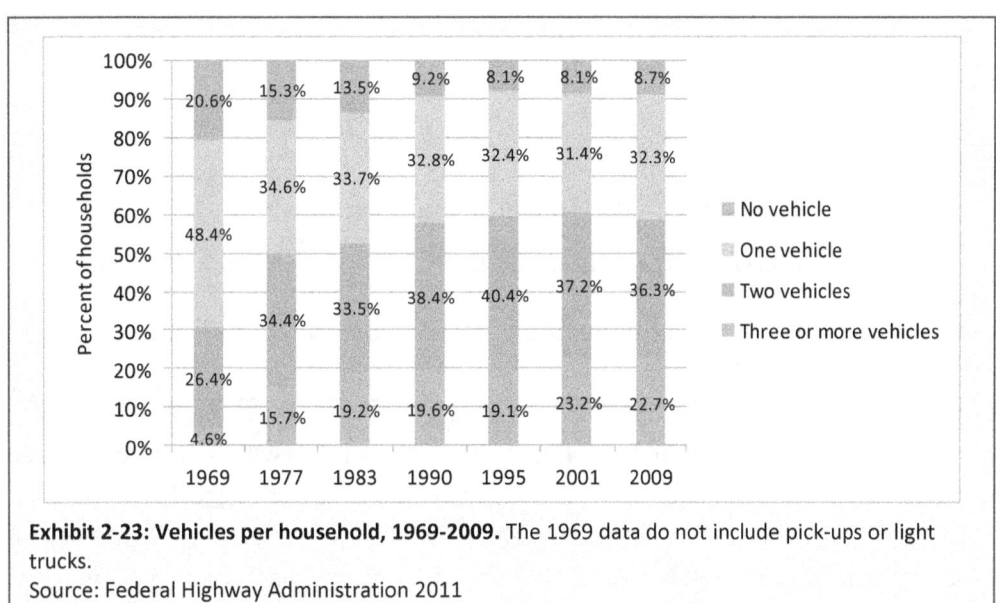

Exhibit 2-23: Vehicles per household, 1969-2009. The 1969 data do not include pick-ups or light trucks.
Source: Federal Highway Administration 2011

In addition to shifting demographics and household finances, much of the increase in vehicle travel is likely a result of outward growth of cities and towns and the roads built to serve the new development. Since the 1950s, development has become more dispersed (Exhibit 2-5). With more households living on large lots far from the economic centers of their communities, workplaces and housing have become more segregated from one another, and distances between everyday destinations have grown. More people used cars for most or all of their trips at the same time that the design of communities changed

---

[73] DiCecio, et al. 2008
[74] DiCecio, et al. 2008. Much of the decline since 2000 can be attributed to a large segment of the population retiring and a sharp decline in teenage workforce participation.

to accommodate more cars. These changes have in turn made it impractical or impossible for many people to get around by any means other than a car, further reinforcing community design that precludes other choices.

2.5.1 Vehicle Travel

While the population roughly doubled between 1950 and 2011, from about 152 million to 312 million people, vehicle travel during this same period increased nearly sixfold, from around 458 billion vehicle miles traveled (VMT)[75] to nearly 3 trillion VMT (Exhibit 2-24).

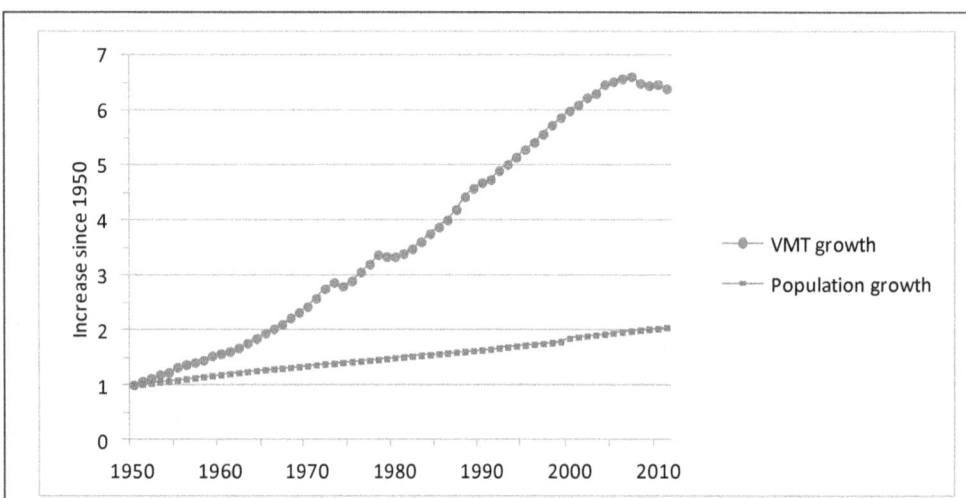

Exhibit 2-24: Growth in VMT and population, 1950-2011. Data are normalized to a 1950 value of 1.0.

Sources: Federal Highway Administration 2010 and 2012; U.S. Census Bureau 2000, 2009, and *State & County QuickFacts* n.d.

The decline in VMT since 2007 (see Exhibit 2-24), the rise in households with no car between 2001 and 2009 (Exhibit 2-23), labor force changes noted above, and other evidence suggest that the decades-long growth of VMT might be slowing. Historically, household income has mirrored VMT, growing at about the same rate—a much higher rate than population growth—and showing similar patterns during economic expansions and contractions. However, since 1997, the trends in household income and VMT have diverged. From 1970 to 1997, VMT grew at 3.0 percent per year, while household income grew at 3.2 percent. From 1997 to 2005, VMT grew at 2.0 percent per year, compared to 3.2 percent for household income.[76] It is unclear why growth in VMT appears to have leveled off or whether the trend will continue, particularly if the rate of economic growth increases. However, research suggests that travel demand might have reached a saturation point as drivers are unwilling to devote more time to

---

[75] Vehicles include passenger cars; other two-axle, four-tire vehicles; motorcycles; and buses.
[76] Memmott 2007

travel, infrastructure improvements no longer allow substantial speed increases, and the marginal benefits of additional trips or travel to additional destinations are not worth the marginal cost.[77]

VMT depends on trip lengths, trip frequencies, and people's choices about how to get around.[78] The increase in VMT thus is related to how close people live to where they work and take care of daily activities, and whether they have viable alternatives to driving. Data from the National Household Travel Survey suggest the number of car trips taken per person could explain more of the rise in VMT than the length of those trips, at least until the mid-1990s. The annual number of vehicle trips taken per household increased from 1,396 in 1969 to a maximum of 2,321 in 1995, falling to 2,068 in 2009, while the average vehicle trip length was relatively constant (Exhibit 2-25).[79] While average household size declined in the United States from 3.2 people in 1969 to 2.5 people in 2009, the average number of vehicles per household increased from 1.2 to 1.9 in the same period. The number of workers and the number of licensed drivers in the United States also more than doubled, explaining in part the increase in the number of trips taken and the increase in VMT.[80]

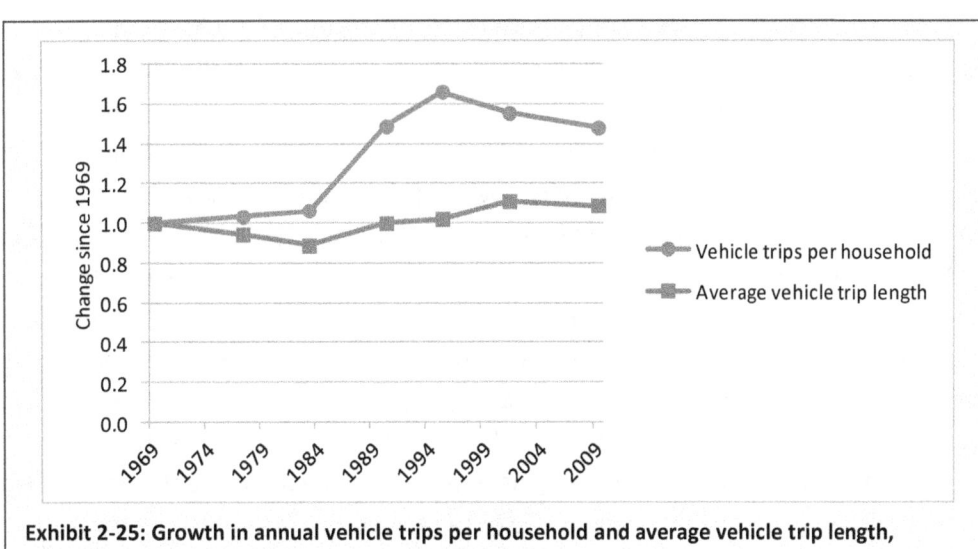

Exhibit 2-25: Growth in annual vehicle trips per household and average vehicle trip length, 1969-2009. Data are normalized to a 1969 value of 1.0.
Source: Federal Highway Administration 2011

2.5.2 Induced Travel

Many researchers have studied the possible effects of additional road capacity on VMT. *Induced travel* is a term for traffic growth produced by the addition of road capacity. The theory behind induced travel is

---

[77] Millard-Ball and Schipper 2011
[78] Ewing and Cervero 2010
[79] The percentages of trips taken by personal vehicle and by public transit were also relatively constant, although data were available only from 1990 to 2009. The percentage of trips made by foot varied between 1990 and 2009 but did not show a consistent trend over time. Interpretation of results for alternative forms of transportation is also complicated by a change in the survey to explicitly ask respondents about walk trips.
[80] Federal Highway Administration 2011

that of supply and demand. Adding road capacity (supply) reduces the cost of vehicle travel by reducing the costs associated with travel time. When cost goes down, demand goes up. As travel time and monetary costs fall, people travel more.

Different types of induced travel would be expected to occur in the short and long terms. In the short term, additional road capacity can lead to people making more trips, increasing trip length, changing the time of travel, or switching from transit or carpools to driving alone because of improved traffic conditions. In the long term, reduced travel costs can encourage more dispersed land use patterns that, in turn, could increase trip lengths and vehicle dependency, leading to a permanent increase in travel demand.

Researchers have long attempted to determine whether additional road capacity actually causes increases in VMT and how great the effect might be.[81] A 2002 review of studies found strong evidence that additional transportation capacity induces travel, both from short-term increases in demand and from long-term changes in land use.[82] Evidence suggests a 10 percent increase in lane miles increases VMT by 3 to 6 percent, although a more recent study found more modest effects.[83] Most recently, a 2011 analysis of city-traffic and road data for 228 metropolitan areas in the United States between 1983 and 2003 found that a 10 percent increase in lane miles on interstate highways was associated with a 10 percent increase in VMT.[84] For other major roads, the increase in VMT was slightly less. Results of this study suggest that the primary cause of the increase was additional truck traffic and an increase in individual VMT rather than population growth in areas with new highways or diversion of traffic from other routes. A study contrary to most results reported no evidence of induced travel.[85] However, this research evaluated trips taken rather than total miles traveled, and this inconsistency could be explained if individual VMT growth is due less to people taking additional trips than to people choosing to travel longer distances to accomplish the same goal or people switching from another form of travel to a car. In fact, evidence suggests that as travel speeds improve, people are willing to travel farther and often do. Average travel time per person per year has remained relatively constant in spite of large increases in road capacity.[86] Although the magnitude of the effect remains uncertain, most evidence suggests that additional road capacity does lead to increased VMT.

Researchers have also investigated whether additional road capacity can cause increases in other types of development. A review of the literature on highway-induced development shows that new highways have little effect on the total amount of development that occurs in a metropolitan area. However, they can have large effects on the location of the development that occurs.[87] Highways tend to drive

---

[81] Handy 2005

[82] Noland and Lem 2002

[83] Cervero 2003

[84] Duranton and Turner 2011

[85] Mokhtarian, et al. 2002

[86] Metz 2008

[87] Ewing 2008

development to areas where they provide improved accessibility, favoring locations outside of central cities and leading to conversion of undeveloped land.

2.5.3   Transit, Walking, and Bicycling

Since the middle of the last century, relatively few people have gotten to work by using public transportation, carpooling, bicycling, or walking. Over most of the last 30 years, these forms of commuting have declined steadily in most areas, with the exception of some cities and towns. Census data on mode choice from 1980 to 2010 show a downward trend in carpooling and walking to work (Exhibit 2-26). Public transportation use also declined each decade from 1980 to 2000 but showed a slight increase in 2010 over 2000 levels. Biking has remained the commuting choice for a relatively constant but very small proportion of all workers. Census data before 1980 did not distinguish between driving alone and carpooling or include information on motorcycle or bicycle use. However, in 1960, 12 percent of people usually took public transportation to work, 2.5 times the percentage in 2010, and 10 percent of people usually walked to work, 3.5 times the percentage in 2010.[88]

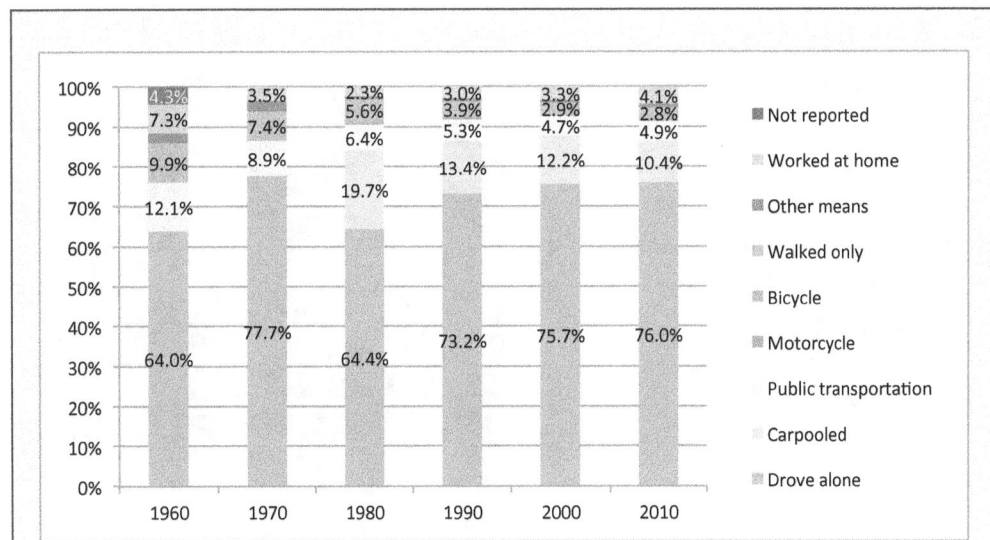

Exhibit 2-26: Means of transportation to work, 1960 to 2010. Data for 1960 and 1970 group carpooling and driving alone together and do not include bicycle or motorcycle use. Data for 1960 only include a "not reported" category.
Sources: Davis, Diegel, and Boundy 2012, Table 8.16 and U.S. Census Bureau, *Means of Transportation to Work for the U.S.* n.d.

Although nationally only a small percentage of people bike or walk to work, there is considerable variation across the country. In 2009, the top metropolitan areas for the percentage of workers who commute by bicycle included Corvallis, Oregon (9.3 percent); Eugene-Springfield, Oregon (6.0 percent); Fort Collins-Loveland, Colorado (5.6 percent); Boulder, Colorado (5.4 percent); and Missoula, Montana (5.0 percent). Each of these cities has more than eight times as many people commuting by bike as the

---

[88] U.S. Census Bureau, *Means of Transportation to Work for the U.S.* n.d.

national average of 0.6 percent.[89] A similar pattern is apparent with walking to work. In 2009, the top metropolitan areas for the percentage of workers who walk to work included Ithaca, New York (15.1 percent); Corvallis, Oregon (11.2 percent); Ames, Iowa (10.4 percent); Champaign-Urbana, Illinois (9.0 percent); and Manhattan, Kansas (8.5 percent), all well exceeding the national average of 2.9 percent.[90]

Commuting by public transportation and carpooling shows some racial and ethnic differences. In 2009, white workers were most likely to drive alone (83.5 percent) and least likely to use public transportation (3.2 percent). Black workers used public transportation at more than three times the rate of white workers and showed a commensurate decrease in the percentage that drove alone. Asians, Hispanics, and other races were also more likely than white workers to use public transportation and carpool. The rate of walking to work was relatively constant across racial and ethnic groups, ranging from 2.8 to 4.4 percent.[91] These differences are likely due at least in part to differences in the availability of public transportation and carpooling to people depending on where they live. They are also likely due in part to differences in car ownership rates among racial and ethnic groups. In 2011, the percentages of households without a car were 6.6 percent for whites (not Hispanic or Latino), 20.2 percent for blacks or African-Americans (not Hispanic or Latino), 11.7 percent for Asians, and 12.8 percent for Hispanics or Latinos (of any race).[92]

While adults are less likely to walk to work, children are also less likely to walk to school than in past decades. In 1969, nearly half of students in kindergarten through eighth grade walked to school. By 2009, that figure had fallen to 13 percent.[93] During that time, the percentage of students who rode a school bus stayed relatively constant, while the percentage of students who were driven to school increased. Walking or biking to school

Exhibit 2-27: Bike lane near Filter Square, Philadelphia.
Students and workers take advantage of a dedicated bike lane to get exercise on the commute to work and school.
Photo source: Kyle Gradinger via flickr.com

---

[89] U.S. Census Bureau, *Commuting in the United States: 2009* 2011
[90] U.S. Census Bureau, *Commuting in the United States: 2009* 2011
[91] U.S. Census Bureau, *Commuting in the United States: 2009* 2011
[92] U.S. Census Bureau, *American FactFinder* n.d.
[93] McDonald, et al. 2011

correlates with overall levels of physical activity for school children,[94] and there is some evidence that walking or biking to school can also improve measures of physical fitness.[95] Section 3.7 discusses more fully how the built environment affects activity levels, obesity, and chronic disease.

Commuting captures a lot of the public's attention because congestion tends to peak during rush hour, and people generally have relatively little ability to avoid it because of limited control over when and where they travel. However, the National Household Travel Survey[96] shows the importance of *non-work* trips to overall VMT; they accounted for nearly three quarters of all miles traveled in 2009. These data also show that people tend to use personal vehicles more often for commuting than for other types of travel, including social and recreational trips and travel to church. For example, the data show that for 2009, a personal vehicle was used for 91 percent of trips to "earn a living" but only 82 percent of non-work trips,[97] while walking was used for 3 percent of trips to earn a living but 12 percent of non-work trips. However, in terms of miles of travel, differences in mode of transportation between work and non-work trips were not as large.[98]

2.6   Future Trends

2.6.1   Projected Population Growth

The U.S. population is projected to grow 42 percent between 2010 and 2050, from 310 million to 439 million.[99] These new people will need additional housing and infrastructure. Researchers estimate that between 2005 and 2050, the United States will need 42 percent more, or 52 million, new housing units. In addition, 37 million units will likely be built to replace existing homes.[100] Together, the number of new and replacement units projected to be built in this time is equivalent to about two-thirds of the 132 million housing units that existed in 2011.[101] Researchers also estimate that between 2005 and 2050, the amount of nonresidential space will grow by about 60 percent to 160 billion square feet, and about 130 billion square feet of nonresidential space will be rebuilt, some structures more than once.[102] These projected trends present an opportunity to improve the environmental performance of our built environment. Where and how we build new housing and infrastructure needed to accommodate projected population growth will have important environmental impacts (as discussed in Chapter 4).

---

[94] Faulkner, et al. 2009
[95] Lubans, et al. 2011
[96] Federal Highway Administration, *National Household Travel Survey* n.d.
[97] Non-work trips include the categories family/personal business, school/church, social and recreational, and other.
[98] Results were calculated based on data collected using the National Household Travel Survey Data Extraction Tool (Federal Highway Administration, *Data Extraction and Visualization Prototypes* n.d.).
[99] Vincent and Velkoff 2010
[100] Ewing, Bartholomew, et al. 2008
[101] U.S. Census Bureau, *State & County QuickFacts* n.d.
[102] Ewing, Bartholomew, et al. 2008

2.6.2   Projected Land Conversion

Based on population projections and current development patterns, one study projects the amount of urban land[103] in the contiguous United States to more than double between 2000 and 2050, increasing from 3 percent to 8 percent.[104] Overall, researchers project that about 45,700 square miles of forestland in nonurban areas in the United States in 2000, an area approximately the size of Pennsylvania, will be located within an urban area in 2050, and much of this forestland would be lost to buildings, roads, and other infrastructure.[105] However, the distribution of this projected development is not equal across the 48 contiguous states. Researchers project that more than 50 percent of four states will be urban land by 2050: Rhode Island (71 percent), New Jersey (64 percent), Massachusetts (61 percent), and Connecticut (61 percent). Projections for these states also indicate that they will have the greatest percentage of land that is currently forest become part of an urban area by 2050. However, these states together account for only 3,472 square miles of non-urban forestland projected to be subsumed by urban areas between 2000 and 2050. Other states account for a larger absolute area projected to be subsumed by urban areas in this period, including North Carolina (3,375 square miles), Georgia (2,994 square miles), New York (2,630 square miles), Pennsylvania (2,451 square miles), Texas (2,404 square miles), Alabama (2,074 square miles), and South Carolina (2,029 square miles).

The U.S. Forest Service also periodically assesses trends in the nation's renewable resources to project future forest conditions and evaluate the implications of those projections for the ecosystem services forests provide (see Section 3.1.1). The 2010 assessment is based on three scenarios that vary in population and personal income projections (see Section 4.3 for a discussion of scenario planning).[106] It forecasts that between 1997 and 2060, 60 to 86 million acres of rural land (as much as the size of New Mexico) will be developed, and between 24 and 38 million acres of forests (as much as the size of Florida), 19 and 28 million acres of cropland (as much as the size of Tennessee), and 8 and 11 million acres of rangeland (as much as the size of Vermont and New Hampshire together) will be lost.

2.6.3   Projected Changes in Development Trends

The extent to which current development trends continue—and the amount of forestland, wetlands, and other natural areas that will be lost—will depend on many factors. Market preferences, led by changing demographics, will certainly influence the nature of future development. The aging of the U.S. population and immigration are the two main drivers of projected demographic trends. Researchers project the age structure of the U.S. population to shift from 13 percent aged 65 and older in 2010 to 19 percent in 2030. The aging of the baby boom generation[107] will shrink the percentage of the population that is of working age (20-64) from 60 percent in 2010 to 55 percent in 2030, a decrease that

---

[103] Urban land was determined according to the U.S. Census Bureau's 2000 definition.
[104] Nowak and Walton 2005
[105] Nowak and Walton 2005
[106] Wear 2011
[107] People born between 1946 and 1964 are part of the baby boom generation.

would be more significant if not for growth in the number of working-age immigrants.[108] At the same time, demographic projections indicate that the percentage of households with children will decline, from 33 percent in 2000 to 27 percent in 2030. During this time, 83 percent of the net growth in households is projected to consist of households without children.[109] Thus, their needs and the needs of the large baby boom population are expected to dominate the housing market until around 2030, after which a decline in the number of baby boomers is expected to have an equally significant impact on housing needs.[110] A literature review by the National Research Council shows that expected demographic changes are likely to influence housing preferences and travel patterns in ways that could shift real estate development towards more compact growth in which residents travel less to meet their daily needs.[111]

2.7   Summary

The size of virtually every metropolitan area in the United States has expanded dramatically in recent decades. In many places, the rate of land development has far outpaced the rate of population growth, although more recent trends in some areas suggest the pattern could be changing. As the amount of developed land has increased and more and larger homes have been built, buildings, roads, and associated impervious surfaces have grown to serve an increasingly dispersed population. As our communities changed to accommodate cars, the percentages of people taking public transit, walking, and biking declined. Projected population growth and demographic trends suggest that the need for additional development will continue to grow, providing an opportunity to improve the environmental performance of our communities. The next chapter discusses some of the environmental impacts of our current land use and travel behavior.

---

[108] Vincent and Velkoff 2010
[109] Nelson 2009
[110] Myers and Pitkin 2009
[111] National Research Council of the National Academies, *Driving and the Built Environment* 2009

Chapter 3. Environmental Consequences of Trends in Land Use, Buildings, and Vehicle Travel

Although development provides many social and economic benefits, it also comes at a cost. Development has seriously degraded or destroyed many natural areas and caused significant growth in driving—both of which have impacts on the health of critical environmental resources as well as on people. These environmental consequences are particularly important because the effects of development are long lasting and not easily reversible. As a result, the cumulative effects of development decisions are important when considering the long-term health of the environment and communities.

The research summarized in this chapter outlines how significant the impacts have been to the natural environment: to critical habitat for plants and animals, to water resources, to air quality, and maybe most importantly, to the planet itself in the form of global climate change. This chapter also discusses impacts that most directly affect human health apart from the surrounding ecosystem: how the way we build our communities influences levels of physical activity, affects levels of community engagement, and affects the risk of being injured or killed in a traffic crash. Many of the impacts discussed in this chapter help point to potential solutions, including changes in how and where we grow and develop our communities, which are discussed in Chapter 4.

This chapter covers the following environmental and human health impacts of the built environment:

1. Habitat loss, degradation, and fragmentation.
2. Degradation and loss of water resources.
3. Degradation of air quality.
4. Heat island effect.
5. Greenhouse gas emissions and global climate change.
6. Health and safety.

3.1   Habitat Loss, Degradation, and Fragmentation

Habitat loss, degradation, and fragmentation are some of the most direct impacts of development on previously undeveloped land. Construction of new buildings, roads, and other infrastructure often destroys native vegetation. Landscaping in the remaining open space with new lawns and non-native plants often cannot serve the same ecological functions as the vegetation it replaces. In addition to direct habitat loss, development often fragments habitat, isolating patches of natural areas with surrounding buildings and roads.

3.1.1 Effects of Habitat Loss

Habitat destruction and degradation contribute to the endangerment of more than 85 percent of the species listed or formally proposed for listing under the federal Endangered Species Act.[112] This loss is critical because biodiversity is the foundational underpinning of ecosystems; the number and type of plants and animals in an area determines the very structure and function of ecosystems across the planet.[113] Ecosystem functions cover a wide spectrum. The concept of *ecosystem services* describes the benefits that we humans get from the natural world, including its plants, animals, microorganisms, and non-living components. Scientists have categorized ecosystem services into *provisioning services* of things like food, fiber, and fresh water that are necessary for life, *regulating services* like pollination and water purification, *cultural services* like recreation and education, and *supporting services* like photosynthesis and nutrient cycling.[114]

Loss of natural areas affects all of the ecosystem services that they provide. However, not all natural areas are created equal, and the impacts of development depend heavily on what type of ecosystem the area supports. This section highlights two important types of ecosystems in the United States: wetlands and forests.

Loss of Wetlands

Wetlands in particular have garnered considerable attention because of their ecological importance. Wetlands are characterized by soil types, plants, and animals that occur in areas regularly saturated with water. They vary widely depending on local geography, climate, hydrology, and water chemistry, among other factors, and occur in diverse ecosystems all over the world. Although they tend to occur in small patches throughout an ecosystem, the world's largest wetlands include the bottomland hardwood forests, swamps, and marshes of the Mississippi River basin (41,700 square miles) and the marshes and meadows of the Prairie potholes (24,000 square miles).[115]

Wetlands perform invaluable ecological functions. The rate of organic matter production in wetlands is among the

Exhibit 3-1: Lily Pad Lake, Yellowstone National Park, Wyoming. A shallow, open lake covered in water lilies supports a variety of wetland plants along with many species of birds, fish, and other wildlife. Photo source: EPA.

---

[112] Wilcove, et al. 1998
[113] Hassan, Scholes, and Ash 2005
[114] Hassan, Scholes, and Ash 2005
[115] Keddy 2010

highest of any ecosystem. They mitigate flooding and damage from erosion, wind, and waves; facilitate soil formation; provide rich feeding grounds and habitat for water life, waterfowl, mammals, and reptiles; help regulate atmospheric carbon dioxide and methane levels; help maintain the global nitrogen cycle; and improve water quality by removing excess nutrients and some chemical contaminants.[116] These myriad ecosystem services cannot be easily replaced. For example, while constructed levees can serve as storm protection, they cannot provide the host of other ecosystem services that coastal wetlands do.[117]

In the late 1700s, the lower 48 states had an estimated 221 million acres of wetlands.[118] Less than half those acres remain today.[119] The annual wetland losses measured since the 1950s have decreased due to protection under the Clean Water Act and restoration efforts, and wetland acreage even increased between 1998 and 2004 (Exhibit 3-2). However, researchers estimate 13,800 acres of wetlands were lost annually between 2004 and 2009, although the difference between the 2004 and 2009 national estimates was not statistically significant. Development contributed to the loss of 128,570 acres, or 23 percent of the total loss between 2004 and 2009,[120] much of which was offset by wetland reestablishment and creation. However, the success of restoration efforts varies,[121] and this inventory does not include information on the wetlands' quality or condition.

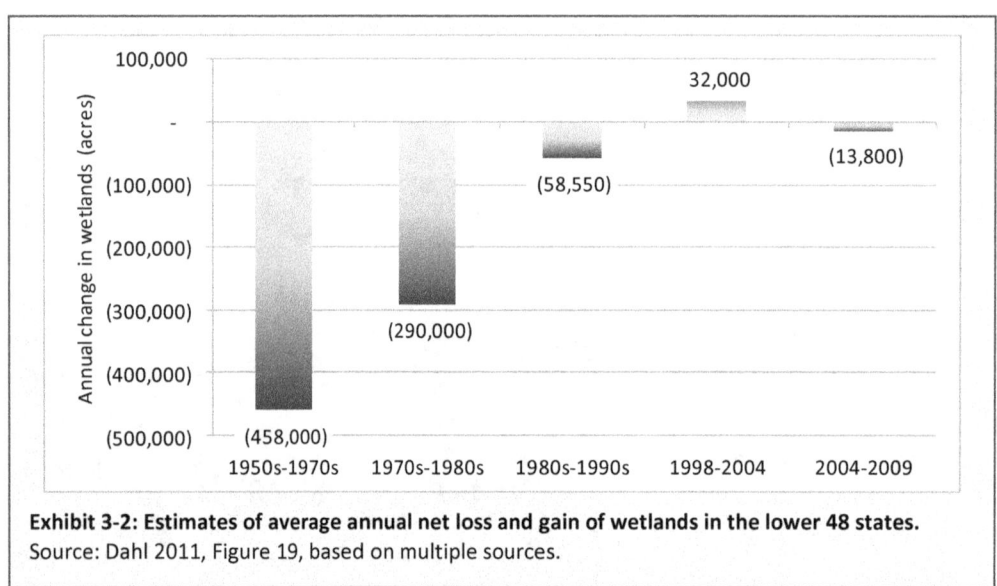

Exhibit 3-2: Estimates of average annual net loss and gain of wetlands in the lower 48 states.
Source: Dahl 2011, Figure 19, based on multiple sources.

In spite of an overall positive trend in national wetland acreage, results are not uniformly encouraging. Coastal wetland restoration is often more difficult and more expensive than interior wetland

---

[116] Keddy 2010

[117] Costanza, et al. 2008

[118] Dahl 1990

[119] Dahl 2011

[120] Tree cultivation was the most significant cause of wetland loss (56 percent), and the remainder (21 percent) was due to wetland conversion to deep, permanent water bodies.

[121] Suding 2011

restoration, so restoration efforts generally offset fewer of the wetland losses along the coasts than in other parts of the country.[122] For example, between 1998 and 2004 when the country gained 32,000 acres of wetlands annually, coastal wetlands along the Gulf of Mexico declined by 61,800 acres annually, and coastal wetlands along the Atlantic Ocean declined by 2,500 acres annually. Nearly one-quarter of the freshwater wetlands lost along the coasts is attributed to urban and rural development.[123]

Loss of Forests

Forests, too, are important ecosystems in the United States. Among the many ecosystem services they provide are helping to regulate the global carbon cycle and hence the rate of global climate change and conserving our soil and water resources.[124] For example, just over half of the nation's water supply originates on forested land.[125]

Most of the forest loss in the United States occurred in the 17th to 19th centuries. Researchers have estimated that forest covered just over 1 billion acres, or 46 percent of the land in the United States, in 1860—a figure that dropped to 759 million acres (34 percent) by 1907, a loss of almost 300 million acres, predominantly to create cropland.[126] Since 1907, forest cover in the United States has been relatively stable. The U.S. Forest Service estimates that in 2007, the United States had 751 million acres of forests.[127]

Although nationally the total amount of forest cover has been relatively stable, some regions have shown significant losses. The northeastern United States has had a great deal of forest regrowth on abandoned cropland, while a lesser amount of regrowth has occurred in Appalachia and around the Great Lakes. However, other parts of the United States experienced significant declines since the early part of the 20th century, including western states (particularly California and Texas) and Florida (Exhibit 3-3).[128] Most of the land newly

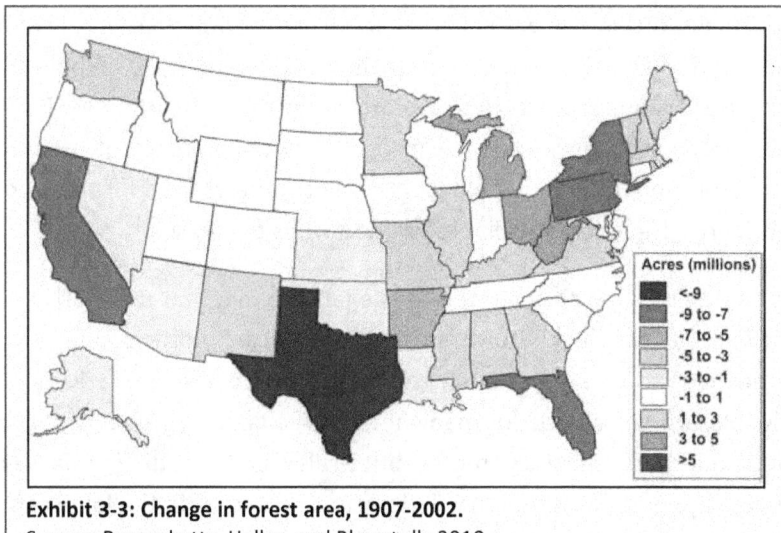

Exhibit 3-3: Change in forest area, 1907-2002.
Source: Ramankutty, Heller, and Rhemtulla 2010

---

[122] Stedman and Dahl 2008
[123] Stedman and Dahl 2008
[124] Hassan, Scholes, and Ash 2005
[125] U.S. Department of Agriculture Forest Service 2009
[126] Smith and Darr 2004
[127] Smith, Miles, et al. 2009
[128] Ramankutty, Heller, and Rhemtulla 2010

developed for urban expansion was previously forested, amounting to 4.7 million acres lost between 1973 and 2000.[129]

National statistics on forest cover also obscure other regional threats beyond outright loss. These threats include changes in forest ownership leading to loss of active forest management, substantial increases in forest fragmentation, and declines in overall forest quality due to factors such as more tree deaths from insect infestation.[130]

3.1.2   Effects of Habitat Degradation

Managed Landscapes

A significant amount of development in the United States since the 1950s has occurred at relatively low densities, with residential buildings surrounded by driveways, sidewalks, patios, lawns, and other managed landscaping. In 2009, the median housing lot size was 0.27 acres, but more than 40 percent of the lots were more than one-half acre.[131] Many of these suburban landscapes support wildlife such as deer, foxes, turtles, and snakes that are rare in more urban areas. However, the widespread replacement of millions of acres of native vegetation with primarily non-native ornamental plants in managed landscapes is a growing problem for the organisms that depend on native plants for food, shelter, and places to rear their young.

The impact of a non-native species depends on its ecological context, and many non-native species provide important ecological benefits.[132,133] One fundamental role of plants in an ecosystem is to create food for herbivores that can transfer their stored energy to higher-level predators. However, homeowners and landscapers have often chosen non-native species for their resistance to insects. In fact, most insect species lack the physiological and behavioral adaptations needed to use non-native plants for food. If ornamental plants cannot serve as food for the same number and diversity of herbivores, the energy available for food webs decreases.[134]

Many studies have documented the negative effect that non-native plants can have on the abundance and diversity of insect herbivores. Many have used moths and butterflies as surrogates for all insect herbivores because information about their host plants is relatively robust and their larvae are especially valuable sources of food for many birds.[135] A comparative study of suburban properties found that the abundance and diversity of moths, butterflies, and breeding birds was positively correlated with the percentage of native grasses, wildflowers, and shrubs in the landscape. For example, there were four times as many moth and butterfly larvae and three times as many moth and butterfly species on properties planted with native ornamental plants than on properties with a mix of native and non-native

---

[129] Drummond and Loveland 2010

[130] U.S. Department of Agriculture Forest Service 2011

[131] U.S. Census Bureau, "American housing survey for the United States: 2009" 2010

[132] Davis, et al. 2011

[133] Simberloff 2011

[134] Tallamy and Shropshire 2009

[135] Tallamy and Shropshire 2009

ornamentals.[136] A literature review documents that native woody plants support three times as many moth and butterfly species as introduced plants, while native woody plants used as ornamentals support 14 times as many moth and butterfly species as non-native ornamentals.[137] In a study of planted garden plots, non-native plants supported significantly fewer caterpillars of significantly fewer species than native plants, even when the non-natives were close relatives of native host plants.[138] Researchers found similar results in a desert ecosystem. A study of neighborhoods in central Arizona showed that native desert bird species exist in greater numbers in neighborhoods with desert landscaping designs and in neighborhoods closer to large desert areas.[139]

Suburban lawns and landscaped areas have other ecological effects. Construction activity for development tends to compact soil, reducing the ability of air and water to move through it. These changes ultimately reduce soil biodiversity by reducing soil microbial biomass, enzymatic activity, soil fauna, and ground flora.[140] Reduced infiltration also limits the amount of precipitation the land can absorb (see Section 3.3), increases risk of flooding, and reduces ground water recharge.[141] One study showed that residential construction activity in north central Florida reduced the infiltration rates in yards covered with turf by 70 to 99 percent relative to nearby natural forest.[142] Another study in Pennsylvania found that the infiltration rate of lawns of more recently constructed homes (since 2000) was 69 percent lower than for lawns of older homes, and 73 percent lower than for agricultural sites, possibly due to changes in construction practices and/or changes over time as vegetation becomes more established.[143] Lawn and landscaping irrigation can also be a major drain on water resources.[144] Fertilizer, if

Exhibit 3-4: Lawn and landscaping. Managed parks and residential lawns can create large areas with little ecological function that require large amounts of water, chemicals, and energy for mowing.
Photo source: this lyre lark (derya) via flickr.com

[136] Burghardt, Tallamy, and Shriver 2008
[137] Tallamy and Shropshire 2009
[138] Burghardt, Tallamy, and Philips, et al. 2010
[139] Lerman and Warren 2011
[140] Nawaz, Bourrie, and Trolard 2013
[141] Gregory, et al. 2006
[142] Gregory, et al. 2006
[143] Woltemade 2010
[144] Haley, Dukes, and Miller 2007

applied when not needed, can lead to excess phosphate and nitrate in local water bodies.[145] Many lawns and landscapes are also treated with insecticides and herbicides that can eradicate native plants and insects. Finally, mowing uses fossil fuels.[146]

Developed areas, even those developed at relatively low densities, thus fundamentally change the food and other resources available to species in an ecosystem. For nearly all plants and animals, species diversity declines with increases in the amount of impervious surface, road density, time since development, human population density, and building density.[147] Animal species that do well in urban environments tend to be generalists that can adapt to available food resources, and they tend to have a set of common behavioral characteristics: they are less wary of people, they are more aggressive towards members of their own species, and they have altered seasonality (e.g., hibernation behavior).[148] The species composition of both plants and animals within cities tends to be more similar than would be expected based on species found in nearby natural areas due in part to the similar pressures wildlife faces in developed areas and the dominance of non-native species.[149,150]

Invasive Species

Researchers estimate approximately 50,000 non-native plant and animal species have been introduced into the United States, and about 5,000 of these have become established in natural ecosystems, most replacing several native species.[151] Some established, non-native species spread widely or quickly and become *invasive*, causing ecological or economic harm or harming human health.[152] Invasive species are at least in part responsible for the listing of about 42 percent of the species that are threatened or endangered.[153] An analysis of nearly 300 publications on the impacts of invasive species across the world concluded that invasive species tend to harm species and community health, especially for plants.[154]

A study of the change in plant species in Marion County, Indiana, illustrates the problem on a smaller scale. Researchers compared historical records of wild plant species from before 1940 with the species found in surveys conducted between 1996 and 2009. Although the overall number of species remained relatively constant, the number of native plant species declined by 2.4 per year. Researchers found 10 percent fewer native species than existed in the county less than a century before. Many high-value wetland plants were lost and were replaced with invasive shrubs that escaped from cultivation.[155]

---

[145] Brown and Froemke 2012
[146] Hostetler and Main 2010
[147] Pickett, et al. 2011
[148] Pickett, et al. 2011
[149] Pickett, et al. 2011
[150] McKinney 2006
[151] Pimentel, Zuniga, and Morrison 2005
[152] National Invasive Species Council n.d.
[153] Pimentel, Zuniga, and Morrison 2005
[154] Pysek, Jarosik, et al. 2012
[155] Dolan, Moore, and Stephens 2011

Invasive species threaten more than ecosystems. They can also directly affect human health. For example, kudzu infestations in Madison County, Georgia, increase nitrogen cycling in soils, which can increase emissions of nitric oxide, an ozone precursor, by more than 100 percent.[156] A review of the ecological and human health effects of invasive species documents a range of impacts, including:

Exhibit 3-5: English Ivy is native to Europe and Western Asia. In the United States, it is a common landscape plant but also an invasive species. Uncontrolled, the vine can overtake stands of trees. Eventually the weight of the vine can topple trees or kill them by blocking sun from the tree's leaves. It can also completely cover a forest floor, outcompeting other understory plants.
Photo source: English Ivy Nancy Fraley, USDI National Park Service, Bugwood.org

- **Human disease**—For example, the tiger mosquito is a vector for pathogens that cause dengue fever, yellow fever, and malaria.
- **Toxins in human food**—For example, honey made exclusively from the herb Salvation Jane (*Echium plantagineum*) contains toxic levels of pyrrolizidine, which can cause liver damage.
- **Allergies**—For example, pampas grass causes pollen allergies.
- **Injury**—For example, leaves of common cordgrass (native to the U.S. Atlantic coast but invasive on the U.S. Pacific coast) can cut skin, and the rugosa rose creates thorny thickets.
- **Contamination**—For example, the droppings of excessive numbers of Canada geese can contaminate soil and water.[157]

3.1.3   Effects of Habitat Fragmentation

Landscape modification, whether for agriculture or development, not only destroys native vegetation in the area modified, but also harms what native vegetation remains because of nearby increased land-use intensity.[158] A 2007 literature review shows landscape modification and habitat fragmentation negatively affect virtually all taxonomic groups, including birds, mammals, reptiles, amphibians, invertebrates, and plants, and is a severe threat to global biodiversity.[159] The edges of forests, meadows, and other natural areas can have different characteristics than the more central areas. The amount of sunlight and moisture, temperature, humidity, wind speed, and soil nutrients all differ, which leads to

---

[156] Hickman, et al. 2010
[157] Pysek and Richardson 2010
[158] Fischer and Lindenmayer 2007
[159] Fischer and Lindenmayer 2007

differences in the species of plants and animals found there and their patterns of competition, predation, and parasitism. In general, these changes harm native ecosystems,[160] particularly as habitat becomes ever more fragmented and the proportion of it with these effects increases. Development is a common cause of habitat fragmentation. A study of how forest cover varies with different levels of development in New England suggests that development generally begins in a few small patches. As development increases, the number of communities increases, but only up to a certain level, beyond which they begin to coalesce into larger towns and cities. However, fragmentation of forested land continues increasing even beyond this point, possibly due to increasing road density.[161]

The ecological effects of habitat fragmentation have been widely studied. A review of the literature shows that while the total amount of habitat and its quality have a greater and better documented effect on the number of plant and animal species in an area than habitat fragmentation,[162] the effects of fragmentation are wide ranging and significant. A quantitative review of studies shows that habitat fragmentation decreases the genetic diversity of plant populations,[163] which can reduce their ability to adapt to changing conditions, raising their risk of extinction.[164] Fragmentation often decreases population size and increases population density, reducing availability of resources (e.g., food and nesting sites) and leading to home-range overlap and elevated relatedness within populations. All of these changes affect how members of a single species interact among themselves, including mating patterns, often reducing reproductive success. These changes also affect how different species interact with each other, influencing predator-prey, host-parasite, and competitor interactions.[165] Animals can have difficulty moving around to find food or mates and escape from predators. Habitat fragmentation also disrupts large-scale animal movements such as migrations or range changes that could be necessary to adapt to climate change.[166]

Research suggests habitat fragmentation could also directly affect human health through its effects on disease vectors and their hosts. For example, as fragmentation increases, mammalian species diversity decreases, and some tick hosts decline or disappear while mice, which are relatively good reservoirs for the bacterium that causes Lyme disease, tend to dominate. In such areas, modeling suggests that human exposure to Lyme disease would be higher.[167]

---

[160] Fischer and Lindenmayer 2007
[161] Coles, et al. 2010
[162] Di Giulio, Hoderegger, and Tobias 2009
[163] Aguilar, et al. 2008
[164] Spielman, Brook, and Frankham 2004
[165] Banks, et al. 2007
[166] Fischer and Lindenmayer 2007
[167] LoGiudice, et al. 2003

3.1.4   Effects of Roads: Combined Effects of Habitat Loss, Degradation, and Fragmentation

The direct and indirect ecological effects of roads are so significant that their study has emerged as a specific field called *road ecology*.[168] Roads, from local streets to highways, are a large component of the built environment, and they contribute to habitat loss, degradation, and fragmentation.

Roads have numerous impacts on the environment. Precipitation landing on roads can pick up heat and pollutants and travel quickly to water bodies, affecting water quality and causing erosion and subsequent changes to stream channels instead of replenishing ground water where it lands (see Section 3.2). A whole host of chemical pollutants arise from road construction, maintenance, and use, including pesticides, deicing salts, hydrocarbons, asbestos, lead, cadmium, copper, carbon monoxide, nitrogen oxides (NO_X), volatile organic compounds (VOCs), sulfur dioxide, particulates, methane, benzene, butadiene, and formaldehyde. Noise near roadways can be significant. Finally, roadsides tend to be windier, hotter, dryer, sunnier, and dustier than the surrounding natural habitat. The microclimatic effects of even narrow roads determine the plants and soil macroinvertebrates that can survive there. Even birds, amphibians, reptiles, and mammals are affected, contributing to a decline in overall species richness near roadways.[169]

Roads affect wildlife in several ways. First, roads directly destroy habitat.[170] Second, roads contribute significantly to habitat fragmentation[171] and block movement for many animals. Finding food, mates, and breeding

Exhibit 3-6: Road through natural area. Vermont Route 17 is a heavily traveled road that bisects Camels Hump State Forest, creating a hazard for wildlife and fragmenting habitat.
Photo source: EPA

sites can be more difficult or even impossible. These limitations affect the ability of populations to breed with one another, reducing the overall genetic diversity.[172] Third, roads facilitate the spread of non-native and invasive species. Travel along roads helps to disperse seeds, and disturbed areas along roadsides provide long corridors of uninterrupted habitat in which weeds can thrive with little competition from woody plants.[173] Finally, roads are responsible for the deaths of large numbers of

---

[168] Forman, Sperling, et al. 2003
[169] Coffin 2007
[170] Coffin 2007
[171] Coffin 2007
[172] Di Giulio, Hoderegger, and Tobias 2009
[173] Coffin 2007

animals [174] (and humans; see Section 3.7.3). Animals that move slowly or that regularly need to cross a road are particularly affected. For other species, the road is an appealing travel corridor, leading to more animals being killed by vehicles. For example, in areas with heavy snow cover, wildlife often favor plowed roads for travel, increasing their risk of being hit by a vehicle. The Federal Highway Administration estimated that there are 1 to 2 million collisions between vehicles and large animals per year.[175] These collisions cause about 200 human deaths and 26,000 human injuries per year, and they almost always kill the animal involved. Road mortality is a major cause of death for 21 animals on the federal threatened or endangered species list, including the Florida panther, red wolf, American crocodile, California tiger salamander, and Florida scrub jay.[176]

Many studies have found that species density tends to increase as distance from roads increases because many species cannot survive along road edges.[177] Researchers have attempted to define and quantify the zone of land around roads that the road system directly affects ecologically. A study of a Massachusetts suburban highway found that this road-effect zone tends to be asymmetric and variable along its length, but in general, the effects of the factors studied extended more than the length of a football field or more than 328 feet (100 meters) from the road.[178] Some effects occurred more than 0.62 miles (1 kilometer) from the road. For example, moose preferred this area for travel, while grassland birds avoided it. Overall, the impact zone averaged about 0.4 miles (600 meters) wide. As noted in Section 2.3.1, another analysis found that more than 50 percent of all land in the United States would fall within this zone.[179] However, an attempt to estimate the amount of land in the United States directly affected ecologically by roads after accounting for the asymmetric and variable shape of the road-effect zone concluded that this zone covered 20 percent of the land area in the United States as of 2000.[180]

3.2   Land Contamination

Past industrial activity has left a legacy of soil and water pollution at former industrialized sites. Thousands of these potentially contaminated properties, or brownfields,[181] are located in densely populated neighborhoods, often near places where residents gather and children play. Many of these sites are near rivers that once served as valuable transportation corridors. The juxtaposition of toxic

---

[174] Coffin 2007

[175] Federal Highway Administration 2008

[176] Federal Highway Administration 2008

[177] Coffin 2007

[178] Forman and Deblinger 2000

[179] Riitters and Wickham 2003

[180] Forman 2000

[181] Brownfields are defined as "real property, the expansion, redevelopment, or reuse of which may be complicated by the presence or potential presence of a hazardous substance, pollutant, or contaminant" in Public Law 107-118 (H.R. 2869) "Small Business Liability Relief and Brownfields Revitalization Act" signed into law January 11, 2002.

chemicals, human activity, and sensitive environmental habitats can lead to a range of problems, including compromised human and environmental health.

Reliable information on the number of brownfields in the United States is scarce. A 2008 survey of 188 U.S. mayors yielded an estimate of nearly 25,000 brownfield sites among their cities.[182] In 2004, EPA estimated that between 235,000 and 355,000 sites in the United States were contaminated with hazardous waste and petroleum products.[183] An analysis of more than 1,400 New York and Texas properties in state voluntary

Exhibit 3-7: Weirton Steel, West Virginia. As the steel industry declined in much of the country, many mills were completely or partially abandoned, leaving behind large, often polluted sites. Because many towns were built around steel mills, these sites are often in a central location with easy rail, highway, and/or river access, making them important locations for redevelopment in many communities.
Photo source: Bob Jagendorf via flickr.com

cleanup programs suggested that commercial areas have as many potentially contaminated sites as former industrial areas.[184] The study also revealed that areas that have experienced more recent rapid economic growth have as many potentially contaminated sites as older industrial areas. Another study estimated brownfield acreage in two cities based on official lists of contaminated properties and land use history that suggested a high probability of contamination.[185] The researchers estimated that Atlanta has 3,244 acres of brownfields out of 84,750 total acres, covering almost 4 percent of its land area. Out of 52,458 total acres, Cleveland has 3,701 acres of brownfields covering more than 7 percent of its land area. Similarly, state, local, and federal inventories indicate that 1,027 properties in the city of Milwaukee are brownfields, constituting 7.5 percent of its land area.[186]

Poor and minority neighborhoods often have a disproportionately high number of brownfield properties. For example, a study of brownfields in the Detroit region found that census block groups located within a half-mile of a brownfield were 58 percent African-American with a median income of $34,177, while block groups located more than a half-mile from a brownfield were 21 percent African-American with a median income of $55,687. The effects of income were independent of the effects of

---

[182] United States Conference of Mayors 2008

[183] EPA, *Cleaning up the Nation's Waste Sites* 2004. This estimate includes Superfund program sites, Resource Conservation and Recovery Act Corrective Action sites, Underground Storage Tank sites, Department of Defense installations, Department of Energy sites, other civilian federal agency sites, and state and private party sites in mandatory cleanup and brownfields programs.

[184] Page and Berger 2006

[185] Leigh and Coffin 2005

[186] McCarthy 2009

race.[187] A similar study of Milwaukee found that census tracts with an above-average percentage of African-Americans had more than 25 percent more brownfields per square mile than census tracts with a below-average percentage of African-Americans; census tracts with an above-average percentage of Hispanics had twice the number of brownfields per square mile as census tracts with a below-average percentage of Hispanics; and census tracts with an above-average percentage of people below the poverty level had nearly three times as many brownfields per square mile as census tracts with a below-average percentage of people below the poverty level.[188]

3.3   Degradation and Loss of Water Resources

Most developed land in the United States was originally grassland, prairie, or forest.[189] The replacement of these natural ecosystems with buildings, roads, and other infrastructure has several impacts on water resources and water quality. EPA collects reports from states on the results of water body assessments in the Watershed Assessment, Tracking & Environmental Results (WATERS) database. Exhibit 3-8 shows the probable source of impairment for different types of assessed water bodies.

Most sources of impairment are directly or indirectly related to the built environment. However, the category labeled "urban-related runoff/stormwater" is the source most directly connected to the amount of impervious cover on the landscape and thus most directly influenced by where and how we build our communities. Urban-related stormwater is thought to be responsible for the impairment of 858,186 acres of lakes, reservoirs, and ponds and 51,548 miles of rivers and streams, among other types of assessed waters (see Exhibit 3-8). These estimates are likely low because the majority of U.S. waters have not been assessed: only 45 percent of lakes, reservoirs, and ponds; 27 percent of rivers and streams; and 37 percent of bays and estuaries have been assessed. In addition, stormwater discharges come from numerous sources and are episodic and diffuse, making them difficult to identify. Thus, stormwater is thought to be underreported as a source of impairment for many assessed waters.[190] In fact, "other," "unknown," and "unspecified point source" are among the largest categories of probable source groups.

The hydrology of most urban systems is highly modified due to development and flood prevention. Storm drains, water mains, wastewater sewers, and other water infrastructure control the movement of water on the surface and underground throughout many developed areas.[191] Even the natural stream network is often forced into underground pipes that are fed by storm drains.[192] The natural hydrologic

---

[187] Lee and Mohai 2011
[188] McCarthy 2009
[189] National Research Council of the National Academies, *Urban Stormwater Management in the U.S.* 2009
[190] National Research Council of the National Academies, *Urban Stormwater Management in the U.S.* 2009
[191] Price 2011
[192] National Research Council of the National Academies, *Urban Stormwater Management in the U.S.* 2009

system has a set of consistent responses to such changes that affect its hydrology, geomorphology, stream pollution and nutrient levels, and aquatic life.[193]

Probable Source Group	Size of Assessed Waters With Probable Sources of Impairments							
	Rivers and Streams (Miles)	Lakes, Reservoirs, and Ponds (Acres)	Bays and Estuaries (Square Miles)	Coastal Shoreline (Miles)	Ocean and Near Coastal (Square Miles)	Wetlands (Acres)	Great Lakes Shoreline (Miles)	Great Lakes Open Water (Square Miles)
Agriculture	123,669	1,821,113	3,028	142		369,637	536	4,373
Aquaculture	318	4,620	0	5				
Atmospheric Deposition	101,015	4,804,255	7,530	388	777	200,741	3,400	53,270
Commercial Harbor and Port Activities		109,240	470	20	0			
Construction	13,522	319,470	16	23	3	13,971		
Ground Water Loadings/Withdrawals	178	98,032	158			3,045		
Habitat Alterations (not Directly Related to Hydromodification)	32,415	360,800	2,057			8,017		
Hydromodification	58,915	907,651	2,513	136	4	109,024	7	
Industrial	14,200	224,021	3,760	127	4	195,840	31	
Land Application/Waste Sites			1					
Land Application/Waste Sites/Tanks	8,277	77,106	52	12		1,634	31	
Legacy/Historical Pollutants	4,889	769,768	1,469	8			427	13,991
Military Bases	42	2,436						
Municipal Discharges/Sewage	50,813	813,637	4,406	326	116	464,479	120	
Natural/Wildlife	51,396	1,355,736	4,225	100	1	563,032	0	
Other	10,179	867,056	3,630	34		65,874		25
Recreation and Tourism (Non-Boating)	1,741	107,766	0	20	4	787		
Recreational Boating and Marinas	132	126,413	1,053	36	9			
Resource Extraction	26,264	563,235	1,292	12		94,719		
Silviculture (Forestry)	19,444	242,583	0			2		
Spills/Dumping	2,420	194,422	26	13	2			
Unknown	87,600	3,199,313	5,447	427	560	707,582	141	310
Unspecified Nonpoint Source	46,985	759,087	2,607	120	7	63,901	8	
Urban-Related Runoff/Stormwater	51,548	858,186	1,877	270	452	2	2	13,867
Total Impaired	**515,141**	**13,039,283**	**21,575**	**7,263**	**1,016**	**1,107,261**	**4,353**	**53,270**
Total Assessed	971,796	18,945,401	32,668	9,010	1,984	1,317,011	4,431	53,332
Total Not Assessed	2,561,409	22,720,648	55,122	49,608	52,136	106,382,989	772	7,214

Exhibit 3-8: Probable sources contributing to water quality impairments nationally, 2012. EPA classifies each cause of impairment reported by states into one of the listed causes of impairment for reporting purposes. Blank cells indicate no states reported anything for that source group for that water type group. Cells with zero values indicate one or more states did report that source, but the total reported was less than 0.5. This table reports the most currently available information as of September 2012, which varied by state.
Source: EPA, *Watershed Assessment, Tracking & Environmental ResultS* n.d.

---

[193] Walsh, et al. 2005

3.3.1 Effects of Development on Stream Hydrology

Development causes fundamental changes to the water (or hydrologic) cycle, the pattern of movement of water on, above, and below the earth's surface through processes such as evaporation, precipitation, and infiltration.

The first effects occur once vegetation is removed from the landscape. Removing vegetation decreases the amount of precipitation that returns to the atmosphere through evaporation from the earth's surface and transpiration from plant surfaces (together called *evapotranspiration*). In arid regions of the country where most precipitation is evapotranspired, this reduction alone has a particularly significant impact on the water cycle.[194] However, in developed areas, not only is the amount of vegetation diminished, sometimes significantly or even entirely, but impervious surfaces such as pavement and rooftops increase, and the amount of water that runs off the surface of the land increases.

Increased runoff in turn has several effects on the water cycle. First, it reduces the amount of water that recharges underground water storage areas and moves through subsurface pathways. These ground water channels largely sustain streams, so their reduction can lead to reduced flows during dry periods. Second, increased runoff makes floods more frequent and severe during wet periods, as water that would normally soak into the ground near where it lands instead cannot infiltrate.[195]

Increased runoff is not only a problem of increased water quantity, but also of increased water speed as it flows. An increased quantity of water moves faster, reaches peak flow more quickly after precipitation begins, and flows for a longer period of time—all of which increase erosion and flood risk.[196] In addition, flow characteristics affect aquatic species whose behavior is generally adapted to a particular pattern of flow.[197]

Not only does the amount of runoff entering water bodies increase as impervious cover increases, the temperature of the water tends to increase as well. Many impervious surfaces are dark. They transfer the heat they have absorbed to the water that lands on or runs over them. Warmer water holds less oxygen for aquatic species and can disrupt mating, reproduction, foraging, and predator escape.[198]

Changes in the water cycle due to development can be substantial. For example, in a study of two similar areas in the Piedmont region of North Carolina over five years and seven months, one area remained as forest and agricultural fields while the other was developed as a residential subdivision of 32 single-family homes, which resulted in 53 percent impervious cover. In the area that was developed, the volume of runoff was 68 percent greater, and the ratio of runoff to rainfall was 2.75 times that in the undeveloped area. Much of the difference is accounted for by decreased shallow, subsurface flows in the developed area. One day after most rain events and two days after large rains (more than 50

---

[194] National Research Council of the National Academies, *Urban Stormwater Management in the U.S.* 2009
[195] National Research Council of the National Academies, *Urban Stormwater Management in the U.S.* 2009
[196] National Research Council of the National Academies, *Urban Stormwater Management in the U.S.* 2009
[197] National Research Council of the National Academies, *Urban Stormwater Management in the U.S.* 2009
[198] Frazer 2005

millimeters), the developed area had no stormwater remaining to discharge. The undeveloped area discharged 25 percent of the total precipitation after this period because it had been able to hold the water and discharge it more slowly.[199]

Changes in the water cycle due to development are also widespread. An analysis of streamflow alterations at 2,888 stream monitoring stations across the United States found that 86 percent differed in some way compared to a nearby reference location in a relatively undisturbed area. Moreover, the greater the degree of stream flow alteration, the less able a stream was to support and maintain life based on multiple measures of community health for both fish and macroinvertebrates.[200]

Although many impacts of urbanization are consistent across regions, particular impacts can also be unpredictable. For example, although decreased infiltration often reduces the amount of precipitation that flows through underground pathways, in many urban areas, this baseflow can actually increase. Water distribution pipes, particularly those in older cities that might be in operation beyond their expected useful life, can leak large amounts of water. Irrigated lawns and other areas can also increase baseflow beyond what would occur naturally because underground pipes might carry the water from another watershed.[201]

Regional differences across the country also occur. An analysis of four large hydrologic regions of the United States measured the correlation between 10 hydrologic variables and the proportion of the watershed area that is urban.[202] Exhibit 3-9 shows the statistically significant results, revealing that impacts of development on stream hydrology are diverse and can vary across geographic regions.[203]

Variable	Central	Northwest	Southeast	Southwest
Peak flow	Increased	—	Increased	—
Minimum daily streamflow	Increased	Decreased	—	—
Days per year with no streamflow	—	—	—	Increased
Duration of moderately high flows	—	Decreased	Decreased	—
Flow variability	Increased	Increased	Increased	—

Exhibit 3-9: Effects on streams of increasing development in different regions of the United States. Results were not statistically significant in cells marked with "—".
Source: Poff, Bledsoe, and Cuhaciyan 2006

3.3.2   Effects of Development on Stream Geomorphology

Geomorphology is the arrangement and characteristics of landforms, including stream channels, ridges, hill slopes, and flood plains. Most natural stream channels have a complex structure that creates variation in flow (and thus habitat) along the channel. A riparian zone occupies the transition between

---

[199] Line and White 2007

[200] Carlisle, Wolock, and Meador 2011

[201] Price 2011

[202] The authors used data from the U.S. Geological Survey National Land Cover Database to classify areas into one of three classes: least disturbed, including forests, grasslands, wetlands, open water, and bare rock; agriculture; and urban, including low- and high-intensity mines, residential areas, and other areas planted with grasses such as parks or golf courses.

[203] Poff, Bledsoe, and Cuhaciyan 2006

the stream channel and upland areas. In humid regions, shallow ground water forms flow paths to streams. These areas can support anaerobic microbial activity that provides a sink for inorganic nitrogen and reduces nitrate loads flowing to the stream. In arid regions, riparian areas can be the only areas in the watershed that have sufficient moisture to support the vegetation that provides critical habitat and other ecosystem services.[204] Flood plains, which are periodically inundated, are integral parts of river systems.[205]

Exhibit 3-10: Stream bank erosion from stormwater. Heavy flows have scoured the bank of Whiteley Creek in Garards Fort, Pennsylvania, leaving roots exposed, toppling trees, and increasing the amount of sun falling on the stream, which raises the water temperature.
Photo source: EPA

The hydrologic changes due to development cause several visible geomorphologic changes to stream systems. Increased flow washes out woody debris and uproots trees and other vegetation, which destabilizes stream channels and can harm aquatic habitat. Even if forest regeneration occurs, natural rates rarely compensate for the losses. In many developed areas, forest regeneration cannot occur at all.[206] Erosion can eventually lead to stream bank collapse, which deepens and widens channels and can shift normal flow levels many feet below their original level. [207] Such a change in channel geomorphology can lower ground water levels nearby. With lower ground water levels, subsurface water flows to the channel only through deeper paths that lack the organic material found at the surface. Without this organic material, which supports anaerobic microbial activity that reduces nitrate, additional nutrients flow to the stream.[208] Such hydrologic changes are due to increases in impervious cover in developed areas, but in many cases, people modify stream channels directly, straightening and lining them to increase capacity, reduce flooding risk, and fix the position of the stream.[209]

Several studies have documented changes in channel geomorphology with development,[210] although the relationship is complex. For example, a study of 30 streams in nine U.S. metropolitan areas found that watershed-scale indicators of stream habitat quality were associated with the percentage of

---

[204] National Research Council of the National Academies, *Urban Stormwater Management in the U.S.* 2009
[205] Booth and Bledsoe 2009
[206] Booth and Bledsoe 2009
[207] National Research Council of the National Academies, *Urban Stormwater Management in the U.S.* 2009
[208] National Research Council of the National Academies, *Urban Stormwater Management in the U.S.* 2009
[209] National Research Council of the National Academies, *Urban Stormwater Management in the U.S.* 2009
[210] Jacobson 2011

impervious surface only in some metropolitan areas and environmental settings.[211] This and other studies[212] found that multiple factors influence how stream channels respond to development, including climate, hydrology, slope, vegetation, and the presence of bedrock. In addition, historical alterations of the drainage network for stormwater management, grade control, bank stabilization, and other reasons continue to influence stream channels, complicating interpretation of studies that consider only the most recent alterations.

3.3.3 Effects of Development on Water Pollution and Nutrients

Most small, natural streams have very low levels of toxic chemicals; relatively low levels of dissolved solids such as calcium, nitrate, phosphorus, iron, and sulfur; and relatively low levels of suspended solids such as silt, algae, and organic debris. Development increases their concentrations in water bodies through stormwater runoff, which picks up lawn fertilizer and pesticides, pet waste, trash, pollution from vehicles and pavement materials, and chemicals from industrial and commercial activities. When stormwater infiltrates soil, the soil can immobilize many of the pollutants or absorb them until they are broken down. Unless stormwater soaks into the ground or is otherwise treated, it will transport pollutants it picks up to a nearby water body. Pollution also reaches water bodies directly from the air, either settling to the ground as dust or falling in rain or snow, and from direct discharges (e.g., from leaking sanitary sewer and drinking water distribution systems or industrial wastewater). Common pollutants found in stormwater include trash, sediment, nutrients,[213] bacteria, metals, pesticides, and other chemicals.[214,215]

The National Stormwater Quality Database compiles data collected since the early 1990s from nearly 200 municipalities regulated under EPA's stormwater permit program. All samples include information on land use at the collection site, allowing comparisons of water quality across land use types. Statistical analyses of the data show significant differences in pollution levels across land uses for nearly all pollutants measured:[216]

- **Open space** shows consistently low concentrations of all pollutants and other constituents examined.
- **Residential areas** have the highest concentrations of dissolved and total phosphorus and relatively high concentrations of fecal coliform.

---

[211] Fitzpatrick and Peppler 2010

[212] Poff, Bledsoe, and Cuhaciyan 2006

[213] Nutrients (nitrogen and phosphorus) are necessary for living plants and animals. However, excess quantities in our air and water cause environmental and human health problems, including massive overgrowths of algae that can be toxic to humans and can consume so much oxygen in water bodies that other organisms are unable to survive.

[214] National Research Council of the National Academies, *Urban Stormwater Management in the U.S.* 2009

[215] Polycyclic aromatic hydrocarbons are chemicals that could potentially cause health problems at high concentrations. More information about them is available at
www.atsdr.cdc.gov/toxfaqs/tf.asp?id=121&tid=25#information.

[216] National Research Council of the National Academies, *Urban Stormwater Management in the U.S.* 2009

Exhibit 3-11: Road salt can contaminate surface and ground water.
Rain or melting snow will carry road salt dropped from a truck in Dickerson, Maryland directly into the adjacent stream where the resulting water chemistry changes can harm fish and wildlife. Road salt can also contaminate drinking water sources and pose health risks to people.
Photo source: EPA

- **Highway drainage** has the highest concentrations of total suspended solids, chemical oxygen demand (an indirect measure of the amount of organic compounds present), oil and grease, and ammonia. Roads and parking lots can account for as much as 70 percent of the total impervious cover in the most urban areas and can easily capture pollution from vehicles.

An analysis of stream-bed sediment at 98 sites across seven metropolitan regions found contaminants at all sites, and the concentration tended to increase with increasing development of the area around the site.[217] Nevertheless, the patterns of pollution are quite irregular across developed areas, so correlations between measures of development such as the percentage of impervious cover are generally weak.[218] A study of 15 streams near Melbourne, Australia, suggests that a more important variable than the percentage of impervious cover and one with a stronger correlation with pollutant patterns is the amount of impervious area that is directly connected to a stream by pipes or lined drains, also known as *effective impervious area.*[219] Precipitation falling on this type of impervious area is carried directly to streams without an opportunity to infiltrate into the ground.

Prior land use history can help explain why correlations between levels of development and levels of pollution are weak. For example, a study of six different metropolitan areas of the United States found that insecticide concentrations increased with increasing development in all six areas. However, results were less consistent for nutrients, sediment, sulfate, and chloride. In places where development occurred on what had been predominantly forestland or in other naturally vegetated areas, measures of increasing development were generally associated with increasing chemical concentrations. The effects of development were less clear in places where it occurred on land that was already stressed from agriculture, large-scale movement and storage of water, or inflow of relatively saline ground water.[220] A later analysis showed that measurements of nitrogen, phosphorous, chloride, and pesticides not only

---

[217] Moran, et al. 2012
[218] Booth and Bledsoe 2009
[219] Hatt, et al. 2004
[220] Sprague, et al. 2007

tend to increase as naturally vegetated areas are developed, but they also become more variable as water quality fluctuates with seasonal activities such as road salting.[221]

Although correlations can be difficult to detect along a gradient of development, differences between developed and undeveloped areas are prevalent and can be striking. For example, a study of copper, lead, and zinc in samples from storm events in five California watersheds between 2000 and 2005 found that their mean concentrations during storm events and their total loadings per watershed area were significantly greater—often by more than an order of magnitude—in developed than in undeveloped watersheds. The differences were likely largely attributable to stormwater runoff from industrial areas in the developed watersheds.[222] Even strictly residential development can have large impacts. The study discussed in Section 3.3.1 of two similar areas, one undeveloped and one developed as a residential subdivision in North Carolina, found that the amount of sediment exported from the developed area was 95 percent greater and the amount of nitrogen and phosphorous exported was 66 to 88 percent greater.[223]

Development affects not only surface water, but ground water as well. Many studies of ground water quality do not distinguish between industrial, commercial, and residential land uses. However, ground water in developed areas tends to show levels of major ions (e.g., calcium, chloride, nitrate, sodium, and sulfate), pesticides, VOCs (e.g., from degreasers, use in dry cleaning, or use in septic systems), and trace elements (e.g., boron, copper iron, and manganese) at concentrations well above levels found in undeveloped areas.[224] For example, a study of how land use affects water quality of an aquifer in east-central Minnesota found that sewered residential and commercial or industrial areas had higher concentrations of total dissolved solids—including calcium, potassium, sulfate, and magnesium—relative to agricultural, unsewered residential, or undeveloped areas.[225] Unsewered residential areas had higher concentrations of boron, chloride, and nitrate. Researchers found VOCs in all samples from commercial or industrial areas and in about half of the samples from sewered residential areas. Samples taken over a four-year period suggested that when an undeveloped area becomes residential, or an unsewered community is sewered, water chemistry changes rapidly. The data also suggest that prior contamination in older residential and commercial areas might be gradually improving due to cleanup and pollution prevention programs.

An assessment of 55 VOCs in about 3,500 domestic and public drinking water wells across the United States found that, before any treatment, about 20 percent of samples covering 90 out of 98 aquifers contained one or more contaminants at concentrations equal to or greater than 0.2 micrograms per liter, although only 1 to 2 percent of samples had concentrations known to be a potential health

---

[221] Beaulieu, Bell, and Coles 2012

[222] Tiefenthaler, Stein, and Schiff 2008

[223] Line and White 2007

[224] Trojan 2005

[225] Trojan, Maloney, et al. 2003. Samples were collected from wells set 1 meter into the aquifer. At least 90 percent of a 100-meter area that drains to each well consisted of a single land use.

concern.[226] More than 50 percent of samples contained levels of contamination at concentrations equal to or greater than 0.02 micrograms per liter, indicating that although concentrations do not yet raise human health concerns, VOCs are widespread in U.S. aquifers. The most frequently detected compounds included solvents, refrigerants, a gasoline additive, and a gasoline component. The amount of developed land was one of several factors commonly associated with detection.

Pesticides are also common contaminants in both ground water and surface streams. A national assessment of untreated water samples from 186 stream sites and 5,047 wells found that 55 percent of shallow ground water and 97 percent of stream samples in developed areas contained pesticides, compared to 29 percent of ground water and 65 percent of stream samples in undeveloped areas.[227] While none of the samples in undeveloped areas exceeded concentrations known to raise human health concerns, 5 percent of ground water and 7 percent of stream samples in developed areas did exceed such levels. About a quarter of pesticide applications in the United States are for non-agricultural use around homes; in gardens, parks, and golf courses; and along roads. Five herbicides and three pesticides generally used for non-agricultural purposes were among the most frequently detected contaminants.

3.3.4   Effects of Development on Aquatic Life

Although all types of water bodies are affected, most research focuses on streams, which often flow through a range of land use types and so allow relatively straightforward assessment of how various levels of development affect aquatic life. The impacts of watershed development, including increased pollution and temperature and altered channels and flow regimes, reduce the quantity, quality, and diversity of stream habitat for aquatic life.[228] Much of the aquatic life in streams is adapted to a particular

Exhibit 3-12: Mayfly. Mayflies live just a few days as adults but spend months or even years as larval nymphs in freshwater streams or ponds. They are a favorite food for fish and other aquatic predators. They require water with a neutral pH and high levels of dissolved oxygen, and they cannot tolerate pollution, making them important water quality indicators.
Photo source: Andy Nelson via flickr.com

---

[226] Zogorski, et al. 2006

[227] Gilliom, et al. 2006. In this study, land classified as urban contains more than 25 percent urban land and less than or equal to 25 percent agricultural land. Land classified as undeveloped contains less than or equal to 5 percent urban land and less than or equal to 25 percent agricultural land. Land classified as agricultural land contains more than 50 percent agricultural land and less than or equal to 5 percent urban land. All other combinations are classified as mixed.

[228] Booth and Bledsoe 2009

type of stream bed and flow pattern and cannot survive changes caused by frequent periods of high flow that scour the stream bed, removing habitat used as a temporary refuge from the high flow. These changes tend to favor certain types of species—those that have high reproductive potential, are opportunistic, eat a wide variety of foods, and tolerate chemicals.[229,230]

An index of biological integrity is often used to capture the net impact on biological communities of multiple, diverse stressors. A study of three streams in the upper Piedmont region of South Carolina experiencing different levels of land development found that, as the amount of land disturbance increased, impervious cover, stormwater runoff, and total suspended solids associated with storm events increased, while habitat and an index of biological integrity declined.[231]

3.3.5 Levels of Development at Which Effects Are Apparent

The level of development at which negative effects first begin to appear has been the focus of much scientific study, and it has been widely debated whether any threshold exists below which effects are not detectable. Several recent studies support the idea that if such a threshold exists, it is at very low levels of development. A study of nine U.S. metropolitan regions concluded that there is little evidence of a threshold of imperviousness below which there are no effects. Stream invertebrates showed negative impacts at the lowest levels of development, and the response appeared to be linear. In areas without (or with little) prior agricultural use, metrics of overall invertebrate health declined 25 to 33 percent relative to background conditions at just 10 percent impervious cover. At only 5 percent impervious cover, the decline was 13 to 23 percent.[232] Analysis of a large biomonitoring data set from Maryland showed declines of 110 out of 238 macroinvertebrate[233] taxa at low levels of impervious cover. Approximately 80 percent of the declining taxa began declining at between 0.5 and 2 percent impervious cover. The remaining 20 percent did so at levels between 2 and 25 percent.[234] Finally, an analysis of data from 357 fish collections in the Etowah River basin of Georgia between 1998 and 2003 found that some species become rare at impervious cover levels as low as 2 percent.[235]

In 1994, researchers first proposed a model describing the relationship between the amount of impervious cover in a watershed and stream health.[236] A review of literature over a decade later found that most research continued to support the model, but this additional work suggested refinements.[237] In the modified model, there are four stream categories based on the degree to which impervious cover affects their quality (Exhibit 3-13). The least affected category of stream, a sensitive stream, has an

---

[229] National Research Council of the National Academies, *Urban Stormwater Management in the U.S.* 2009

[230] King, et al. 2011

[231] Sciera, et al. 2008

[232] Cuffney, et al. 2010

[233] A *macroinvertebrate* is an animal with no backbone that can be seen with the naked eye. Macroinvertebrates include insects, spiders, worms, snails, slugs, and crayfish.

[234] King, et al. 2011

[235] Wenger, et al. 2008

[236] Schueler 1994

[237] Schueler, Fraley-McNeal, and Cappiella 2009

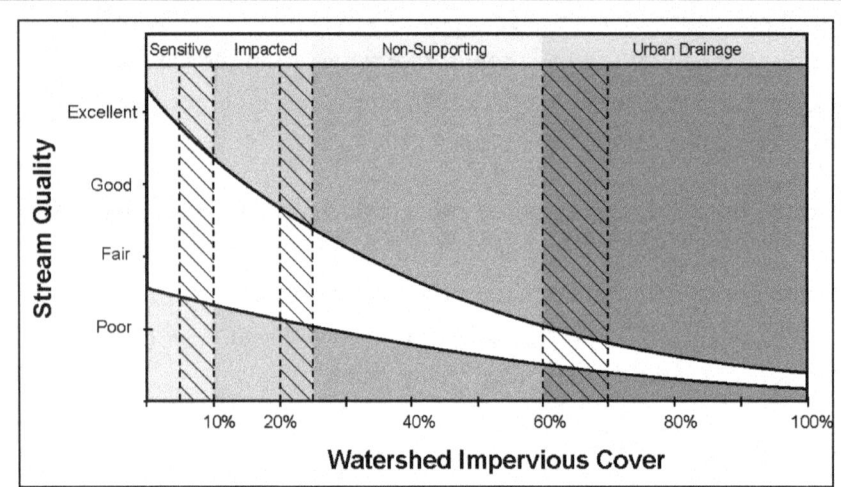

Exhibit 3-13: Impervious cover model. The white cone represents the observed variability in the response of streams to different levels of impervious cover. As the percentage of impervious cover in a watershed increases, stream quality declines. The dashed lines indicate that the transition point between stream classifications is variable. Image source: Schueler, Fraley-McNeal, and Cappiella 2009. Reprinted by permission of the publisher (American Society of Civil Engineers).

impervious cover level of less than 5 to 10 percent. Stream quality with this level of impervious cover is generally fair to excellent, with quality influenced heavily by other characteristics of the watershed such as the amount of forest cover, agriculture, and road density. Stream quality shows the greatest variability at these lowest levels of impervious cover. Streams with impervious cover levels greater than 5 to 10 percent and less than 20 to 25 percent are *impacted* streams that show clear signs of declining stream health. Streams with impervious cover greater than 20 to 25 percent and less than 60 to 70 percent are *nonsupporting* streams that are so degraded they no longer maintain their hydrology, channel stability, habitat, water quality, and/or biological diversity. Finally, streams with impervious cover greater than 60 to 70 percent are classified as *urban drainage,* with quality that is consistently poor. Urban streams might exist solely in storm drains at these levels of imperviousness.

3.3.6   Loss of Water Resources

As water resources are polluted and degraded, they can become unfit for drinking, swimming, fishing, and other uses. However, water resources can also be strained if our use exceeds the available supply. About 86 percent of U.S. households rely on public water supplies for their household use. About one-third of the water from public water supplies comes from ground water, and two-thirds comes from surface water such as lakes and streams.[238] For households that supply their own water, 98 percent rely on ground water.[239] Ground water use can exceed the rate at which precipitation soaking into the ground can replenish it, leading to ground water depletion. Dry wells, reduced amounts of water in streams and lakes, lower water quality, and land subsidence can result.[240] Impervious surfaces created by development[241] and centralized wastewater treatment[242] can decrease rates of ground water

---

[238] Kenny, et al. 2009
[239] Kenny, et al. 2009
[240] U.S. Geological Survey 2003
[241] Frazer 2005
[242] Vaccaro and Olsen 2007

recharge, exacerbating the effects of increased demand on ground water supplies. Ground water depletion has been a concern in the Southwest and High Plains for many years, but increased demand has stressed sources in many other areas of the country, including the Atlantic coastal plain, west-central Florida, and the Gulf coastal plain, among others.[243]

3.4 Degradation of Air Quality

3.4.1 Criteria Air Pollutants

Land use, development, and transportation affect air quality in significant ways. For common air pollutants, EPA has established and regularly reviews National Ambient Air Quality Standards (NAAQS) to protect public health and the environment. In setting or revising primary health-based standards, the Agency considers the effects of poor air quality on at-risk populations such as children and the elderly.[244] EPA has set standards for six principal pollutants (so-called "criteria pollutants"): carbon monoxide, nitrogen dioxide, sulfur dioxide, lead, coarse particulate matter (PM_{10}), fine particulate matter ($PM_{2.5}$), and ozone.[245] VOCs and NO_X are precursors to the formation of ozone.[246]

With the exception of lead, which was phased out of gasoline from 1973 to 1996, motor vehicles contribute to all of these forms of air pollution. Motor vehicles emit pollution through fuel combustion (exhaust) during operation and fuel evaporation during and between periods of operation. Gasoline-powered vehicles are major contributors of pollution from VOCs, NO_X, and carbon monoxide. For diesel vehicles, emissions of NO_X and $PM_{2.5}$ raise the most serious health concerns.[247] Fuel standards and vehicle technology to reduce emissions from cars and other light-duty vehicles have dramatically decreased the amount new cars pollute. Compared with a vehicle sold in 1970, one sold 40 years later emits 99 percent less carbon monoxide, NO_X, particulate matter, and VOCs.[248] Technology to reduce diesel engine emissions was developed much later than technology to reduce gasoline engine emissions, and heavy-duty vehicles tend to have a longer life span before being replaced with newer, less polluting vehicles.[249] Reductions in diesel emissions have therefore been more modest, but declines will grow as older vehicles are retired. EPA and the National Highway Traffic Safety Administration have set increasingly stringent fuel economy standards through model year 2025 to continue improvements in air quality and reductions in greenhouse gas emissions (see Section 3.6).[250]

---

[243] U.S. Geological Survey 2003

[244] EPA, *Our Nation's Air* 2010

[245] EPA, *National Ambient Air Quality Standards* 2012

[246] Sawyer 2010

[247] Sawyer 2010

[248] Sawyer 2010

[249] Sawyer 2010

[250] National Highway Traffic Safety Administration, *CAFE – Fuel Economy* n.d.

Despite already impressive reductions in vehicle emissions since 1970, gains would be even greater if not for the approximately 250 percent increase in VMT since then (see Section 2.5.1).[251] More than 38 percent of national carbon monoxide emissions and 38 percent of nitrogen oxide emissions come from highway vehicles (Exhibit 3-14). Stationary sources like power plants that provide energy to homes, offices, and industries are also major sources of pollution. Fuel combustion for residential, commercial, and industrial uses is responsible for 83 percent of sulfur dioxide emissions and 32 percent of NO_x emissions.[252]

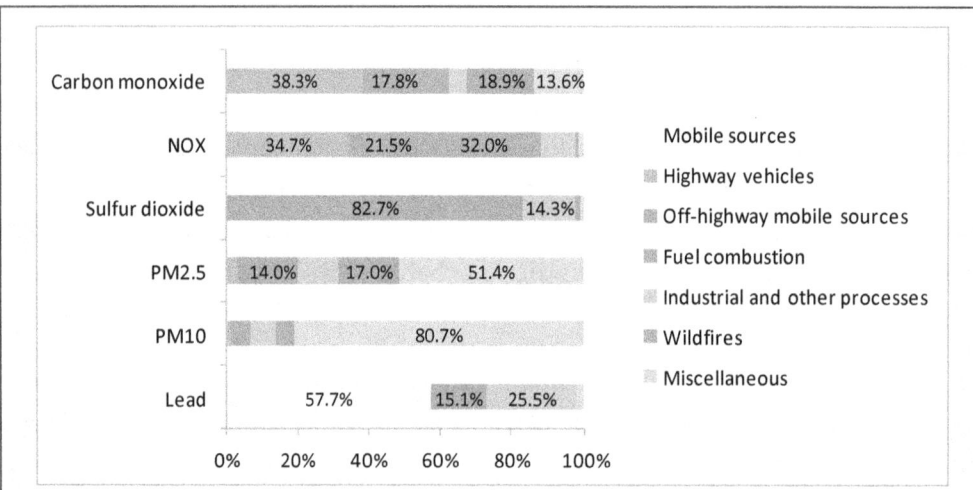

Exhibit 3-14: National emissions by source sector, 2010. Miscellaneous sources include dust from construction and unpaved roads, agriculture, and others. Lead emissions data are from 2008 and combine highway vehicles and off-highway mobile sources into the single category "mobile sources."
Sources: EPA, *Air Emission Sources* 2008 and "1970-2012 average annual emissions" 2012

The amount people drive clearly has a significant bearing on air pollution levels. However, the amount of infrastructure needed to accommodate cars contributes to air pollution regardless of the amount of miles driven. A study that computed the lifecycle emissions of sulfur dioxide and PM_{10} for cars showed that adding parking lot construction and maintenance to the calculations raises emissions by as much as 24 percent and 89 percent, respectively, over calculations excluding these factors.[253] Another study, which reviewed 14 lifecycle assessments of road construction, concluded that the energy used in road construction equals the energy used by traffic on the road for one to two years.[254] In addition, the construction of a lane mile of road produces the equivalent of the annual carbon dioxide emissions of 20 U.S. households.[255]

---

[251] Sawyer 2010

[252] EPA, *Our Nation's Air* 2010

[253] Chester, Horvath, and Madanat 2010

[254] Muench 2010

[255] This calculation is based on the median value of carbon dioxide emissions for road construction projects reviewed in the paper.

Despite progress in cleaning the air, as of 2012, approximately 159 million people—about half the U.S. population—still lived in counties that failed to meet air quality standards, most frequently for ozone and particulate matter.[256] The Centers for Disease Control and Prevention evaluated the race and ethnicity of the people living in counties failing to meet standards for $PM_{2.5}$ and for ozone based on 2006-2008 and 2007-2009 data, respectively (Exhibit 3-15).[257] Asians and Hispanics were most likely to live in counties with poor air quality—more than 2.5 times as likely as non-Hispanic whites to live in a county failing to meet standards for $PM_{2.5}$ and more than 50 percent more likely than whites to live in a county failing to meet standards for ozone. Another analysis of 2005-2007 national air quality monitoring data that ranked all communities according to their air pollution levels found that the proportion of non-Hispanic blacks in communities with the highest levels of $PM_{2.5}$ and ozone is more than twice the proportion in communities with the lowest levels.[258]

Race/Ethnicity	Percent of Population in Nonattainment Counties for $PM_{2.5}$ (2006-2008)	Percent of Population in Nonattainment Counties for Ozone (2007-2009)
White, non-Hispanic	9.7	32.0
Black, non-Hispanic	15.2	40.0
American Indian/Alaska Native	8.2	18.6
Asian	26.2	50.2
Native Hawaiian/Other Pacific Islander	22.0	36.5
Non-Hispanic, multiple races	15.2	36.1
Hispanic	26.6	48.4

Exhibit 3-15: Percentage of racial/ethnic groups living in nonattainment counties for $PM_{2.5}$ and ozone in the United States. Nonattainment counties did not meet the National Ambient Air Quality Standards for the 2006 24-hour $PM_{2.5}$ standard of 35 micrograms per cubic meter from 2006-2008 and the 2008 8-hour ozone standard of 75 parts per billion from 2007-2009.
Source: Yip, et al. 2011

3.4.2   Air Toxics

In addition to criteria air pollutants, EPA regulates hazardous air pollutants, also referred to as *air toxics*. Air toxics are pollutants known or suspected to cause cancer or other serious human health effects or ecosystem damage. Persistent air toxics are of particular concern in aquatic ecosystems, as toxic levels can magnify up the food chain. Compared with the criteria pollutants, less information is available about the health and environmental impacts of current levels of exposure to most individual air toxics.

Although natural sources such as volcanic eruptions and forest fires are responsible for some air toxics releases, most air toxics are released by manmade sources such as cars, trucks, buses, and large, stationary sources such as factories, refineries, and power plants. Some materials used in building construction and some products such as cleaning materials can also cause indoor air exposures to air toxics.[259]

---

[256] EPA, *Summary Nonattainment Area Population Exposure Report* 2012
[257] Yip, et al. 2011
[258] Miranda, et al. 2011
[259] EPA, *About Air Toxics* n.d.

EPA conducts a National-Scale Air Toxics Assessment every three years that estimates the risk of cancer and other serious health effects from inhaling air toxics. In 2011, EPA released its assessment based on 2005 data, the most recent year for which complete information was available.[260] According to the 2005 assessment, mobile, on-road sources like cars, trucks, and buses accounted for 39 percent of total benzene emissions, 38 percent of total naphthalene emissions, 33 percent of total 1,3-butadiene emissions, 21 percent of total formaldehyde emissions, and 8 percent of total acrolein emissions, among other air toxics.[261]

In many places, exposure risk to air toxics varies by race and socioeconomic status. For example, one study in Seattle and Portland, Oregon, compared the race and socioeconomic characteristics of all people living in the cities' urban growth areas with the characteristics of the subset of those people living near a road.[262] Along high-volume roads in both cities, the concentration of African-Americans is two to three times greater than the concentration of these groups in other areas of the city, and the concentration of people living below the poverty level is 1.2 to 1.4 times greater.[263]

3.4.3 Human Health and Environmental Effects of Air Pollution

Air pollutants are associated with numerous public health problems and ecological effects. Exhibit 3-16 shows the wide range of effects of individual air pollutants. Further, many of these pollutants can also have synergistic effects, where their combined effect is greater than or different from the additive effect of the individual pollutants, making reductions in the source of emissions potentially more important than the reduction of individual pollutants through specific control technologies.[264]

Many of these health, environmental, and climate effects occur despite significant improvement in our nation's air quality since 1990.[265] For example, an analysis of the health effects of ground-level ozone and $PM_{2.5}$ based on modeled 2005 concentrations across the United States found that 130,000 to 340,000 premature deaths are attributable to these pollutants annually.[266] The number of life years lost due to ground-level ozone and $PM_{2.5}$ varies by age. People over age 65 lost an estimated 1.1 million life years due to exposure to $PM_{2.5}$ and 36,000 life years due to ozone exposure, approximately 7 percent of total life years lost among this cohort in 2005. The effects of air pollution vary among population subgroups, including people of lower socioeconomic status, people of color, the young, and the elderly, due in part to disproportionate exposure and a higher prevalence of underlying diseases that increase susceptibility to air pollution.[267,268]

---

[260] EPA, *2005 National-Scale Air Toxics Assessment* n.d.

[261] EPA, "Air toxics pie charts" 2011

[262] The study considered people to be living near a road if they lived within 330 feet of a road having at least 100,000 vehicles per day, a zone in which people likely have higher risk of exposure to mobile-source air pollution.

[263] Bae, et al. 2007

[264] Giles, et al. 2011

[265] EPA, *Our Nation's Air* 2012

[266] Fann, et al. 2012

[267] Frumkin 2002

Pollutant	Human Health Effects	Environmental and Climate Effects
Ozone	• Decreases lung function and causes respiratory symptoms such as coughing and shortness of breath. • Aggravates asthma and other lung diseases, leading to increased medication use, hospital admissions, emergency department visits, and premature mortality.	• Damages vegetation by injuring leaves, reducing photosynthesis, impairing reproduction and growth, and decreasing crop yields. • Ozone damage to plants can alter ecosystem structure, reduce biodiversity, and decrease plant uptake of CO_2. • Contributes to the warming of the atmosphere.
Particulate matter	• Short-term exposures can aggravate heart or lung diseases leading to symptoms, increased medication use, hospital admissions, emergency department visits, and premature mortality. • Long-term exposures can lead to the development of heart or lung disease and premature mortality.	• Impairs visibility, harms ecosystem processes, and damages and/or soils structures and property. • Has variable climate impacts depending on particle type. Most particles are reflective and lead to cooling, while some (especially black carbon) absorb energy and lead to warming. • Changes the timing and location of traditional rainfall patterns.
Lead	• Damages the developing nervous system, resulting in IQ loss and impacts on learning, memory, and behavior in children. • Causes cardiovascular and renal effects in adults and early effects related to anemia.	• Harms plants and wildlife, accumulates in soils, and harms both terrestrial and aquatic systems.
Sulfur oxides (SO_x)	• Aggravate asthma, leading to wheezing, chest tightness and shortness of breath, increased medication use, hospital admissions, and emergency department visits. • At very high levels can cause respiratory symptoms in people without lung disease.	• Contribute to the acidification of soil and surface water and mercury methylation in wetland areas. • Injure vegetation and local species losses in aquatic and terrestrial systems. • Contribute to particle formation with associated environmental effects. Sulfate particles contribute to the cooling of the atmosphere.
Nitrogen oxides (NO_x)	• Aggravate lung diseases leading to respiratory symptoms, hospital admissions, and emergency department visits. • Increase susceptibility to respiratory infection.	• Contribute to the acidification and nutrient enrichment (eutrophication, nitrogen saturation) of soil and surface water. • Lead to biodiversity losses. • Affect levels of ozone, particles, and methane with associated environmental and climate effects.
Carbon monoxide	• Reduces the amount of oxygen reaching the body's organs and tissues. • Aggravates heart disease, resulting in chest pain and other symptoms leading to hospital admissions and emergency department visits.	• Contributes to the formation of ozone.
VOCs	• Some cause cancer and other serious health problems. • Contribute to ozone formation with associated health effects.	• Contribute to ozone formation with associated environmental and climate effects. • Contribute to the formation of CO_2 and ozone, greenhouse gases that warm the atmosphere.
Mercury	• Causes liver, kidney, and brain damage and neurological and developmental damage.	• Deposits into rivers, lakes, and oceans where it accumulates in fish, resulting in exposure to humans and wildlife.
Other air toxics	• Cause cancer; immune system damage; and neurological, reproductive, developmental, respiratory, and other health problems. • Some contribute to ozone and particle pollution with associated health effects.	• Harm wildlife and livestock. • Some accumulate in the food chain. • Some contribute to ozone and particle pollution with associated environmental and climate effects.

Exhibit 3-16: Health, environmental, and climate effects of air pollution.
Source: EPA, *Our Nation's Air* 2010

3.4.4 Indoor Sources of Pollution

Although outdoor air pollution receives widespread attention and is the subject of federal regulation, people spend most of their time indoors, so exposure levels from indoor air pollution are of considerable concern as well. Many of the same pollutants in outdoor air are also inside, but people

---

[268] Younger, et al. 2008

indoors can also be exposed to higher concentrations and/or additional indoor sources of pollution including a wide array of chemicals from building materials, household products, and cleaning supplies.[269] People can also be exposed to hazardous chemicals indoors through skin contact and ingestion in addition to inhalation.[270]

The sources and effects of indoor pollutants include:

- **Biological pollutants, including dust mites, fungi, bacteria, and pests (e.g., cockroaches and mice)**—The primary health effects include allergic reactions, from inflammation to asthma. Infections and toxic reactions can also occur.[271]

- **VOCs, i.e., toxic gases that are emitted from certain substances at room temperature. VOCs include benzene, dichlorobenzene, ethylbenzene, chloroform, formaldehyde, methyl tertiary butyl ether, perchloroethylene, tetrachloroethene, toluene, and xylenes**—They can be found in personal care products, cleaning products, paints, pesticides, building materials, and furniture. The primary health effects vary depending on the compound. They include eye and respiratory irritation, rashes, headaches, nausea, vomiting, shortness of breath, and cancer.[272]

- **Asbestos, found mainly in older insulation, but also a wide variety of building materials and products including some vinyl floor tiles, shingles, and heat-resistant fabrics**[273]—The primary health effects are lung cancer, mesothelioma, and (with occupational exposures) asbestosis.[274]

- **Incomplete combustion products of solid fuels, including carbon monoxide and particulate matter (e.g., from wood used for home heating)**—The primary health effects of particulate matter include respiratory irritation, respiratory infections, bronchitis, and lung cancer. Carbon monoxide exposure can cause low birth weight, headaches, nausea, and dizziness.

- **Radon gas**—The primary health effect is lung cancer.[275] Radon is the second leading cause of lung cancer death, after smoking.[276]

- **Polychlorinated biphenyls (PCBs), found in a wide variety of consumer products, including caulking in older buildings, paints, plastics, adhesives, and lubricants**[277]—PCBs have been shown to cause several adverse effects in animals, including cancer and immune, reproductive, nervous, and endocrine system effects. Human studies provide evidence supporting these effects.[278]

- **Polybrominated diphenyl ethers (PBDEs), flame retardants found in a variety of consumer products including upholstery, construction materials, and electrical appliances**[279]—PDBEs are

---

[269] Colbeck and Nasir 2010

[270] Air Force Institute for Environment, Safety, and Occupational Health Risk Analysis n.d.

[271] Perez-Padilla, Schilmann, and Riojas-Rodriguez 2010

[272] Perez-Padilla, Schilmann, and Riojas-Rodriguez 2010

[273] EPA, *Learn About Asbestos* n.d.

[274] Perez-Padilla, Schilmann, and Riojas-Rodriguez 2010

[275] Perez-Padilla, Schilmann, and Riojas-Rodriguez 2010

[276] Al-Zoughool and Krewski 2009

[277] Rudel and Perovich 2009

[278] EPA, *Health Effects of PCBs* n.d.

[279] Rudel and Perovich 2009

not chemically bound and can degrade into particles found in air and house dust.[280,281] Animal research suggests PBDEs can cause neurodevelopmental, kidney, thyroid, and liver toxicity and can disrupt endocrine systems.[282]

The federal government has not established standards for safe levels of most indoor air pollutants.[283] Far less information is available about population exposure levels and health effects from indoor air pollution than for outdoor air pollution. Study design and interpretation are complicated by the difficulty of assessing exposures and effects of chemical and biological mixtures and by variation across buildings resulting from outdoor pollutant concentrations, building ventilation, building materials, relative humidity, temperature, and activities of the building occupants.[284] In addition, the relative scarcity of research for any specific pollutant is due in part to the sheer number of chemicals and potential health effects that require further analysis.[285]

A literature review found that indoor air pollutant exposures have changed a great deal since the 1950s.[286] Exposure to known and suspected carcinogens (e.g., benzene, formaldehyde, asbestos, chloroform, and trichloroethylene) has decreased, as has exposure to recognized toxics such as carbon monoxide, sulfur dioxide, nitrogen dioxides, lead, and mercury. However, indoor levels of endocrine disruptors, chemicals that mimic or block natural hormones, have increased. Known or suspected endocrine disruptors include phthalates found in certain flexible plastics, some flame-retardant chemicals, bisphenol-A, and nonylphenol.

The public health literature on air pollution focuses predominantly on children, who suffer disproportionately from air pollution because their respiratory systems are still

Exhibit 3-17: Baby exploring at home. Children might be more vulnerable to exposure to indoor contaminants because their lungs and metabolic systems are still developing and they breathe and eat more per pound of body weight than adults. Children also play close to the ground where they might consume contaminated dust and tracked-in pollutants, and they might put toys and household objects in their mouths.
Photo source: Ramona Gaukel via stock.xchng

---

[280] Johnson, et al. 2010

[281] Dodson, et al. 2012

[282] EPA, *Technical Fact Sheet* 2012

[283] Bernstein, et al. 2008

[284] Colbeck and Nasir 2010

[285] Weschler 2011

[286] Weschler 2009

developing.[287] A review of research published between 1996 and 2007 that considered the effect of indoor air pollution on the respiratory health of children under the age of five found only a few studies with widely varying methodologies, which limits broad conclusions.[288] However, the results suggest that several indoor air pollutants, including nitrogen dioxide and VOCs, are associated with adverse respiratory effects in young children at levels commonly encountered in developed countries. A review of 21 epidemiologic studies on the association between indoor chemical exposures and respiratory health in infants and children found a relatively consistent association between elevated risk of respiratory or allergic effects and exposure to formaldehyde-emitting materials (e.g., paints, adhesives, insulation, and cabinetry), flexible plastics, and recently painted surfaces.[289] Excess indoor moisture is also associated with adverse health effects. A review of epidemiologic studies found associations between indoor dampness or mold and a range of respiratory and allergic health effects, including asthma, bronchitis, shortness of breath, coughing, respiratory infections, and other upper respiratory tract symptoms.[290]

Similar to the exposure to outdoor air pollution discussed in Sections 3.4.1 and 3.4.2, racial disparities exist in exposure to indoor pollution, although less research has focused on this issue. The socioeconomic status of a household tends to influence several factors that can affect indoor pollution levels, including the size and design of housing, the amount of leakage or air exchange with the outdoors, cooking practices, and the selection of consumer products.[291] Analysis of exposure to 10 VOCs for a sample of the U.S. population aged 20 to 59 found that Mexican-Americans had greater exposure to benzene and 1,4-dichlorobenzene than non-Hispanic whites and blacks. Non-Hispanic blacks had higher exposures than both other groups to chloroform and than Mexican-Americans to tetrachloroethene. The differences were due mainly to associations between race and home location (which determines whether household water is chlorinated), use of air fresheners, and use of dry cleaners.[292] A different analysis of the same dataset showed that race or ethnicity was the strongest determinant of exposure levels among all factors studied. Hispanics and blacks had the highest exposures to benzene, toluene, ethylbenzene, xylenes, methyl tertiary butyl ether (a gasoline additive), and dichlorobenzene. In addition to race, other important risk factors for exposure to one or more of the studied toxics were: living in a home with an attached garage; keeping windows closed year round; smoking; using dry cleaning, stain removers, chlorinated water, moth repellents, and air fresheners; and being exposed to gasoline, fuels, paints, and glues.[293] A study of households in Los Angeles, Houston, and Elizabeth, New Jersey, considered the cancer risks of Hispanics and non-Hispanic whites due to 12 air toxics. For both Hispanics and whites, the cancer risk for nine of the 12 pollutants was greater than the EPA benchmark of 10^{-6}. In both Houston and Elizabeth, Hispanics had a higher combined cancer risk

---

[287] Fuentes-Leonarte, Tenias, and Ballester 2009

[288] Fuentes-Leonarte, Tenias, and Ballester 2009

[289] Mendell 2007

[290] Mendell, et al. 2011

[291] Adamkiewicz, et al. 2011

[292] Wang, et al. 2009

[293] D'Souza, et al. 2009

than whites, mainly due to exposure to *p*-dichlorobenzene, a component of some deodorizers, air fresheners, and moth repellents that does not have a significant outdoor source.[294]

3.5   Heat Island Effect

Not only do impervious surfaces create water quality problems, they also affect the temperature of surrounding areas through what is known as the *heat island effect*. Cities can be as much as 6 to 8 degrees Fahrenheit warmer than outlying areas.[295] The heat island effect is due to two complementary forces: dark pavement and roofs absorb and reflect more of the sun's heat, while the relative scarcity of trees and other vegetation reduces shade and cooling through evapotranspiration.

Increased heat is itself a health hazard, as heat stroke can lead to hospitalization and even death. A review of studies found that while methodological differences hamper efforts to summarize effects, two recent estimates of the effect of temperature on mortality in the United States using identical methods found consistent results for different regions of the country. Both studies found that a 10°F increase in apparent temperature is associated with an approximately 2 percent increase in mortality.[296] However, a study of 107 U.S. communities suggests that the exact relationship will vary by location depending on community characteristics including income, unemployment, population, and the percentage of the population that is urban versus rural.[297] The effects of heat tend disproportionately to affect poor and elderly people.[298]

Increased heat can also lead to other health effects. As temperatures rise, more VOCs are emitted from vehicles, and natural forms of VOCs that are emitted by some tree species increase.[299] Heat also directly increases the rate of ozone formation.[300] In addition, as temperatures rise, people use more air conditioning, which increases air pollution as regional power plants ramp up production and emit more particulate matter, SO_x, NO_x, and air toxics.[301]

3.6   Greenhouse Gas Emissions and Global Climate Change

Greenhouse gases trap heat in the atmosphere and increase the earth's temperature. Several naturally occurring greenhouse gases keep the earth warm enough to support human life. However, human activity is also responsible for a large increase in the amount of greenhouse gases we have in the atmosphere. Carbon dioxide, methane, nitrous oxide, and fluorinated gases are the four main

---

[294] Hun, et al. 2009
[295] Frumkin 2002
[296] Basu 2009
[297] Anderson and Belle 2009
[298] Frumkin 2002
[299] Stone 2008
[300] Frumkin 2002
[301] Frumkin 2002

greenhouse gases emitted from human sources.[302] These greenhouse gas emissions contribute to global climate change, which causes a range of detrimental human health and environmental effects.

3.6.1 Greenhouse Gas Emissions Sources

In 2010, U.S. greenhouse gas emissions totaled 6,821.8 teragrams (or million metric tons) of carbon dioxide (CO_2) equivalent (Tg CO_2 Eq.), an increase of 10.5 percent since 1990.[303] Emissions fluctuate annually due in part to economic conditions, energy prices, and weather. Longer-term changes are due to the size of the population, energy efficiency, and several factors directly related to our built environment, including development patterns and market trends that influence the amount people drive and the size of homes.[304] About 16 percent of total U.S. greenhouse gas emissions in 2010 were offset by the uptake of carbon dioxide in the atmosphere by U.S. sinks, primarily from forests (86 percent) but also from urban trees (9 percent), the management of agricultural soils (4 percent), and landfilled yard trimmings and food scraps (1 percent).[305]

Electricity generation was the largest source of U.S. emissions in 2010, accounting for 34 percent of the total.[306] When emissions from electricity are attributed to an economic sector, industrial activity accounts for the largest portion of total U.S. greenhouse gas emissions, followed closely by transportation (Exhibit 3-18). The residential and commercial sectors together account for just over a third of total U.S. greenhouse gas emissions, with the remainder attributed to agriculture and the U.S.

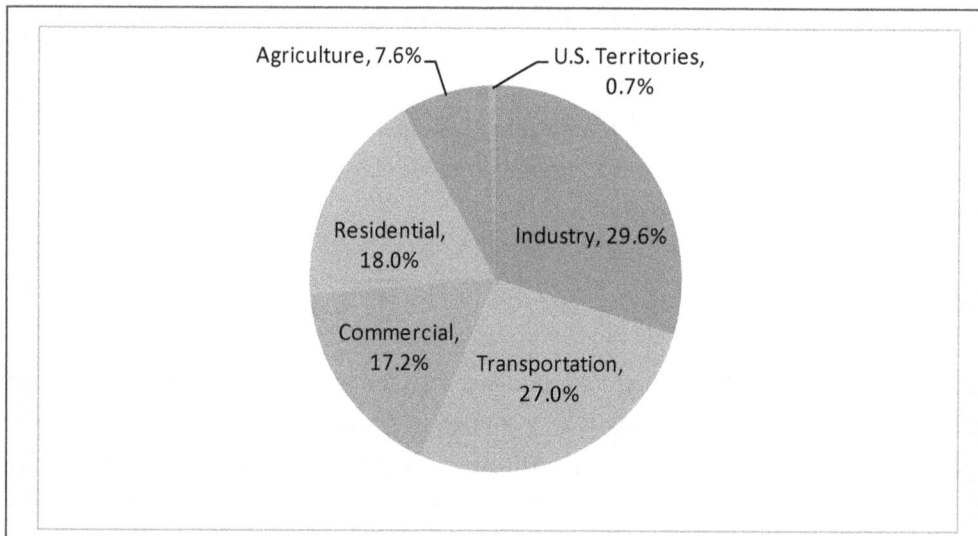

Exhibit 3-18: U.S. greenhouse gas emissions by economic sector including emissions from electricity distributed among sectors, 2010.
Source: EPA, *Inventory of U.S. Greenhouse Gas Emissions and Sinks* 2012

---

[302] EPA, *Greenhouse Gas Emissions* n.d.
[303] EPA, *Inventory of U.S. Greenhouse Gas Emissions and Sinks* 2012
[304] EPA, *Inventory of U.S. Greenhouse Gas Emissions and Sinks* 2012
[305] EPA, *Inventory of U.S. Greenhouse Gas Emissions and Sinks* 2012
[306] EPA, *Inventory of U.S. Greenhouse Gas Emissions and Sinks* 2012

territories.[307] Greenhouse gases from the transportation sector are due to passenger cars (43 percent); freight trucks (22 percent); light-duty trucks, which include sport utility vehicles, pickup trucks, and minivans (19 percent); and commercial aircraft (6 percent).[308]

Transportation emissions increased 19 percent between 1990 and 2010, primarily from carbon dioxide emissions from fossil fuel combustion, but also in part from an increase in hydrofluorocarbons from vehicle air conditioners and refrigerated trucks (Exhibit 3-19).[309] The emissions increase was due primarily to the increase in VMT over this period (see Section 2.5.1), which was partially offset by a slight increase in average fuel economy as older vehicles were removed from the roads.[310] Among new vehicles sold, average fuel economy actually declined between 1990 and 2004 as sales of sport utility vehicles and other light-duty trucks increased. In the 1970s, about one-fifth of new vehicles sold were light-duty trucks. By 2004, they had increased to more than half of the market, although this trend has since begun to reverse.[311] A 4 percent decline in transportation emissions between 2008 and 2009 is due in part to decreased economic activity, particularly the demand for freight transport. As economic activity began to rebound in 2010, transportation emissions increased by 1 percent.[312]

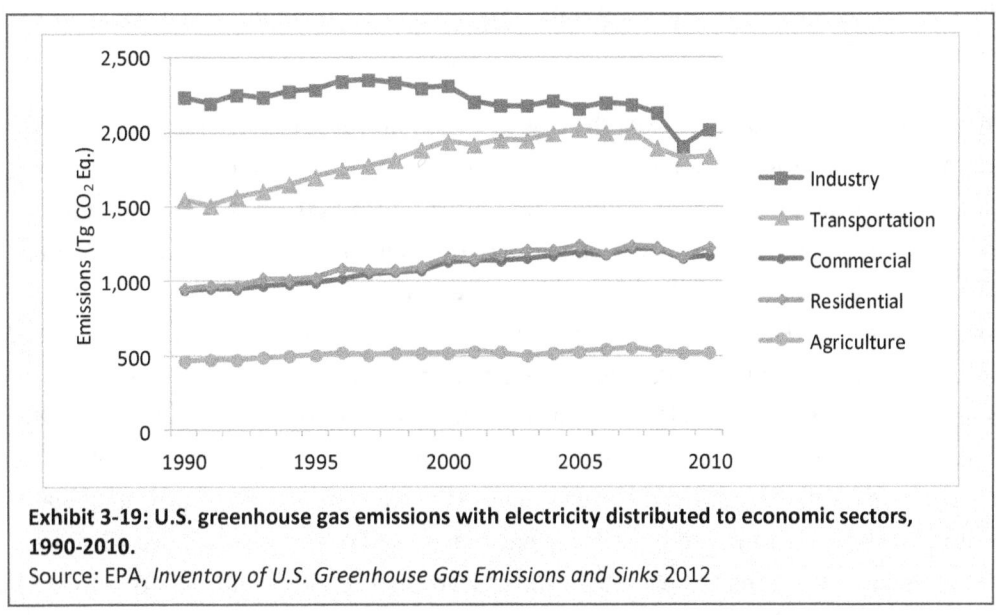

Exhibit 3-19: U.S. greenhouse gas emissions with electricity distributed to economic sectors, 1990-2010.
Source: EPA, *Inventory of U.S. Greenhouse Gas Emissions and Sinks* 2012

Emissions from the residential and commercial sectors also increased relatively steadily between 1990 and 2010: by 25 percent in the commercial sector[313] and 29 percent in the residential sector (see Exhibit 3-19). Both sectors' emissions are due primarily to electricity use, followed by petroleum and natural gas

---

[307] EPA, *Inventory of U.S. Greenhouse Gas Emissions and Sinks* 2012
[308] EPA, *Inventory of U.S. Greenhouse Gas Emissions and Sinks* 2012
[309] EPA, *Inventory of U.S. Greenhouse Gas Emissions and Sinks* 2012
[310] EPA, *Inventory of U.S. Greenhouse Gas Emissions and Sinks* 2012
[311] EPA, *Inventory of U.S. Greenhouse Gas Emissions and Sinks* 2012
[312] EPA, *Inventory of U.S. Greenhouse Gas Emissions and Sinks* 2012
[313] The commercial sector includes landfills and wastewater treatment in these data.

use. Short-term trends in emissions in these sectors are often due to weather conditions that increase the need for heating and/or cooling. In the long term, the emissions trends in these sectors are affected by population growth, population distribution across the country, and changes in the size and energy efficiency of buildings.[314]

3.6.2   Effects of Global Climate Change

How much energy our buildings consume and how much people drive both affect greenhouse gas emissions, making the built environment an important contributor to global climate change. A 2011 report by the National Research Council of the National Academies concluded that global climate change is occurring and is largely due to human activities that lead to heat-trapping greenhouse gas emissions.[315] The scientific academies of over 30 countries[316] and the Intergovernmental Panel on Climate Change, an intergovernmental scientific body, reached the same conclusion based on contributions from thousands of scientists and an extensive review process.[317]

Because climate changes caused by carbon dioxide persist for a very long time, warming will continue well past the end of this century.[318] The 2009 National Climate Assessment found that already-observed changes in the United States include:

- **Temperature**—From 1958 to 2008, the average temperature rose more than 2 degrees Fahrenheit, and precipitation increased by an average of about 5 percent.
- **Rainfall**—Over the past century, the amount of rain falling in the heaviest 1 percent of downpours has increased by an average of about 20 percent, ranging regionally from a 9 percent increase in the Southwest to a 67 percent increase in the Northeast.
- **Extreme weather**—Extreme weather events have become more frequent and intense. Atlantic hurricanes have become stronger and more frequent since the 1980s, although the number of hurricanes making landfall on U.S. soil has not changed substantially. In the Pacific, while storms became stronger over the same time period, they have not become more frequent. Droughts have become more frequent in the West and the Southeast since the 1960s. Heat waves with dangerously high nighttime temperatures have increased over the past 30 to 40 years.
- **Winter storms**—Over the last 50 years, winter storms have shifted northward and are becoming stronger.
- **Wildfires**—In the western United States in particular, wildfires have become more frequent and lasted longer, and wildfire seasons are longer.
- **Sea level**—Sea level along some sections of the U.S. coast rose by as much as 8 inches from 1958 to 2008.

---

[314] EPA, *Inventory of U.S. Greenhouse Gas Emissions and Sinks* 2012
[315] National Research Council of the National Academies 2011
[316] Transportation Research Board of the National Academies 2012
[317] Intergovernmental Panel on Climate Change 2007
[318] National Research Council of the National Academies 2011

- **Sea ice**—Arctic sea ice has shrunk by 3 to 4 percent per decade since 1979.[319]

These impacts are projected to continue to increase over the next century. The length of time over which changes will continue and the severity of the changes depend on the level of greenhouse gas emissions and the climate's sensitivity to those emissions, as well as other factors such as abrupt changes to the climate.[320]

Our built environment affects climate change, but it is also affected by climate change. Changes to the water cycle mean that both floods and

Exhibit 3-20: Flooding of the Monocacy River in Montgomery County, Maryland. Climate change is likely to increase the number and severity of floods in many areas of the country, placing human life at risk while harming ecosystems, dispersing pollutants, and destroying property. Photo source: EPA

droughts are more common, putting lives and property at risk, stressing water infrastructure, and changing the amount and quality of water available for human use. Warmer weather increases energy use for cooling and decreases it for heating, changing the levels and periods of peak demand. Floods, extreme heat, and sea level rise put transportation infrastructure such as roads and rail lines at risk and increase the chances of travel and freight delays and disruptions.[321]

Climate-related changes to the natural environment include major shifts in species' ranges, which can increase the risk of extinction, and migration patterns. Plant growing seasons are changing as well, and pollen seasons for some plants are longer, which affects people allergic to pollen. Changing climate conditions make it easier for some invasive species to take hold (see Section 3.1.2). Plants stressed by drought or heat are more vulnerable to insect pests. Warmer temperatures increase the range of some diseases that affect wildlife. The drier, hotter climate raises the risk of wildfires, which burn habitat.[322]

With the increase in heat waves and extreme heat, the likelihood of illness and death from heat waves also rises (see Section 3.5). Elderly and already-ill people are more vulnerable to adverse health effects from extreme heat, so the growing elderly population and the rising rates of chronic illnesses such as diabetes mean that, without adaptation strategies that help reduce the risk, more people are likely to die from extreme heat. While milder winters could reduce deaths from extreme cold, the number of heat-related deaths is likely to outstrip that reduction, resulting in a net increase in death rates.[323]

---

[319] U.S. Global Change Research Program 2009
[320] U.S. Global Change Research Program 2009
[321] U.S. Global Change Research Program 2009
[322] U.S. Global Change Research Program 2009
[323] U.S. Global Change Research Program 2009

The higher temperatures and increased water vapor attributable to global climate change are projected to increase ozone pollution in already-polluted areas (see Section 3.4.1). By 2050, if current pollution levels remain constant, days on which ozone concentrations are considered unhealthy for everyone are projected to increase by 68 percent—due solely to warming temperatures—in the 50 largest cities in the eastern part of the country.[324]

The increase in extreme events also increases the risk to human life and well-being. Floods can increase sewer overflows, which can contaminate drinking water. Hurricanes and other strong storms can directly threaten human life and can also cause mental health effects such as post-traumatic stress disorder and depression. Wildfires can cause respiratory illnesses in addition to direct harm from burns.[325]

3.7   Other Health and Safety Effects

The built environment's effects on human health extend beyond exposure to air and water pollution or global climate change as already discussed. How we build our communities affects health and safety in several ways. The built environment affects levels of physical activity, obesity, and chronic disease. It also influences our emotional health and the degree of engagement in our communities. Finally, how we design our streets and towns affects the likelihood of being hurt or killed in a vehicle crash.

3.7.1   Activity Levels, Obesity, and Chronic Disease

How we build our cities, towns, and suburbs affects the amount of time people spend in their cars and the opportunity, practicality, and necessity of physical activity in the course of meeting daily needs. These factors can influence overall physical activity levels and therefore the risk of obesity and chronic disease, including heart disease, hypertension, stroke, high cholesterol, osteoarthritis, gall bladder disease, type 2 diabetes, and some cancers.[326,327] These public health threats are of increasing concern as the number of overweight and obese Americans grows. For adults aged 20 to 74, the prevalence of obesity more than doubled between the late 1970s and 2007-2008.[328] As of 2008, 68 percent of adults[329] and 18 percent of children ages 2-9[330] were overweight or obese.

A considerable body of work exists on the association between the built environment and obesity. Broad conclusions are hard to reach due to differences across studies in the features of the built environment that were studied. Nevertheless, a 2010 literature review of 63 papers concluded that the degree of land use mix and county-level measures of sprawl are among the most common metrics used and therefore

---

[324] U.S. Global Change Research Program 2009
[325] U.S. Global Change Research Program 2009
[326] Frank, et al. 2006
[327] Frumkin 2002
[328] Ogden and Carroll 2010
[329] Flegal, et al. 2010
[330] Ogden and Carroll 2010

allow the most robust conclusions, namely that both likely affect the incidence of obesity.[331] For example, one study found that a 10 percent increase in how evenly square footage is distributed across land use types (residential, public, and commercial) was associated with a nearly sixfold increase in walking for transportation and a 25 percent reduction in the prevalence of overweight and obese adults, while controlling for sociodemographic factors.[332] Another comprehensive review in 2012 also noted the challenges of summarizing the literature on the built environment and physical activity given methodological shortcomings and differences across studies.[333] The main finding of this review was that while data are lacking to resolve whether the built environment *determines* levels of physical activity and/or obesity, nearly 90 percent of studies found a positive association, suggesting that the built environment is one of the many factors that could play a role in how much people exercise and levels of obesity.

Some subpopulations are more affected by obesity and related chronic disease than others.[334] Mexican-American women and children; Native Americans; Pacific Islanders; and poor black men, white women, and children have disproportionate levels of obesity at all ages.[335] A possible contributing explanatory factor is that these groups tend to have fewer and poorer-quality recreational facilities in their neighborhoods. A national analysis showed that as the number of minorities and people of lower socioeconomic status increases in an area, the number of physical activity and recreational facilities often decreases.[336] Similarly, an analysis of the Los Angeles metropolitan area shows that Latinos, African-Americans, and low-income groups are more likely to live in neighborhoods with smaller parks and higher population density.[337] In addition, the public park spaces that exist tend to receive less maintenance and are often perceived as unsafe. A study of 685 neighborhoods in North Carolina, New York, and Maryland looked at the availability of public recreational facilities and parks and residents' race

Exhibit 3-21: Meridian Hill Park in Washington, D.C. Parks and other recreational facilities can provide opportunities for physical activity close to home.
Photo source: EPA

[331] Feng, et al. 2010

[332] Li, et al. 2008

[333] Ferdinand, et al. 2012

[334] Ferdinand, et al. 2012

[335] Wang and Beydoun 2007

[336] Gordon-Larson, et al. 2006. Recreational and physical activity facilities included schools, recreation centers, youth centers, parks, golf courses, instructional studios, sporting and recreational camps, swimming pools, and athletic clubs, among others.

[337] Sister, Wolch, and Wilson 2010

and socioeconomic status. Hispanic/black and Hispanic neighborhoods were seven and nine times, respectively, less likely than white neighborhoods to have a facility. Black and racially mixed areas were two and three times, respectively, less likely than white neighborhoods to have a facility in the area. Parks in these neighborhoods were generally much more equitably distributed than other types of recreational facilities, although their size and quality were not assessed.[338]

Overall, available research suggests that among recreational facilities and resources, public parks tend to be the most equitably distributed.[339] For example, researchers investigated how local park access, walkable neighborhoods, and both together are related to the percentage of the population consisting of Latinos, African-Americans, and children in Phoenix, Arizona. Walkability was measured based on housing density, connectivity, and land use diversity. The most walkable census block groups with the best park access had fewer children, but counter to predictions, they also had more Latinos and African-Americans.[340] However, these areas also had higher crime rates, and parks tended to be smaller. Concern for personal safety and limited options within parks might therefore reduce physical activity in these neighborhoods regardless of available facilities.

In spite of the findings that subpopulations have different levels of opportunity for physical activity in their neighborhoods, relatively little research has focused on the association between the built environment and actual physical activity levels for different subgroups, particularly poor and minority communities.[341] However, a 2002 national survey of 3,600 households with children aged 9 through 13 suggests that the built environment does likely play a role in the availability of opportunities for and participation in physical activity. The survey found that while 13 percent of white parents reported that lack of opportunities in the area was a barrier to their child's participation in physical activities, more than 30 percent of black and Hispanic parents reported the same obstacle.[342]

3.7.2 Emotional Health and Community Engagement

A built environment with homes separated from businesses and jobs leaves people few options but to spend large amounts of time in their cars to get to work and meet their daily needs. Some people like to commute because they value the time spent commuting as a transition between work and home life.[343] However, many people spend more time commuting than they would like,[344] which limits the amount of time and energy they can devote to other activities. A built environment that requires or encourages people to spend a significant amount of time in their cars reduces opportunities for informal, spontaneous social interactions with neighbors and acquaintances, shifts activities away from public

---

[338] Moore, et al. 2008
[339] Moore, et al. 2008
[340] Cutts, et al. 2009
[341] Ferdinand, et al. 2012
[342] Duke, Huhman, and Heitzler 2003
[343] Ory, et al. 2004
[344] Redmond and Mokhtarian 2001

spaces such as playgrounds, and can disrupt community life as people age and must move to find different housing options to suit their needs and incomes at different life stages.[345]

Such observations relate to the concept of *social capital*, the benefit that people get emotionally and physically from interpersonal relationships and the broader benefit to communities when its members participate in political organizations, charitable activities, community organizations, and group recreational activities.[346] A review of the literature shows that neighborhoods with more walkable streets, more public space, and a diverse mix of land uses are associated with improved social capital. Automobile dependence, lack of public spaces, and low density tend to be associated with reduced social capital.[347] A study of three New Hampshire communities had similar results. This study looked at the effect walkability had on residents' levels of social capital.[348] The communities differed in the number of destinations residents reported being able to walk to, such as a friend's house, the post office, or stores. Residents of the more walkable communities reported higher levels of all measures of social capital, including trusting neighbors "a lot," participating in community projects, visits from friends at home, volunteer activities, and attendance at club meetings.

3.7.3   Vehicle Crashes

How much people drive and their travel speed are the most important factors that determine their exposure to the risk of injury or death from vehicle crashes, as well as the risk to others. Although absolute risk estimates vary widely, all studies that have examined the relationship between speed and fatality risk report that risk increases with vehicle speed.[349] Development patterns and the design of our communities influence both how much and how fast people drive.[350] The width of travel lanes and shoulders, measures to slow traffic, and even the presence of street trees influence the risk of injury or death from vehicle crashes.

In 2010, 32,885 people died in the United States in vehicle crashes, including vehicle occupants and pedestrians,[351] down from 50,894 in 1966.[352] Many different factors are undoubtedly responsible for the decline, from safer cars and seatbelt laws to tougher laws against drunk driving.[353] The safety gains over this period would have been greater but for the increase in VMT. While the number of fatalities per number of people declined nearly 60 percent over this period, from 2.6 to 1.1 per 10,000 residents, vehicle fatalities per 100 million VMT declined 80 percent, from 5.5 to 1.1 (Exhibit 3-22).

---

[345] Frumkin 2006
[346] Jackson 2003
[347] Frumkin 2006
[348] Rogers, et al. 2011
[349] Rosen, Stigson, and Sander 2011
[350] Ewing and Dumbaugh 2009
[351] National Highway Traffic Safety Administration, *FARS Data Tables* n.d.
[352] National Highway Traffic Safety Administration 2001
[353] Frumkin 2006

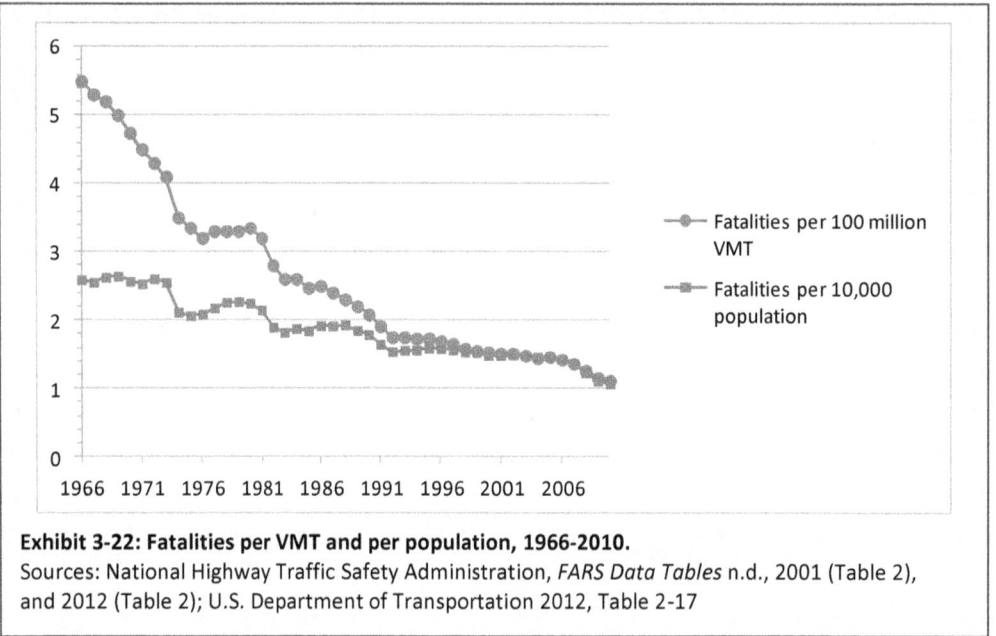

Exhibit 3-22: Fatalities per VMT and per population, 1966-2010.
Sources: National Highway Traffic Safety Administration, *FARS Data Tables* n.d., 2001 (Table 2), and 2012 (Table 2); U.S. Department of Transportation 2012, Table 2-17

In spite of an overall downward trend in traffic fatalities, they are nevertheless a leading cause of death for many cohorts. In 2007, motor vehicle crashes were the leading cause of death for people aged 3 through 33, excluding children aged 6, 7, and 10.[354] Car crashes are the third leading cause of death in terms of years of life lost given the young age of so many car crash victims and the number of years they would have been expected to live if they had not died in a car crash. Only cancer and heart disease are responsible for more years of life lost.[355] High as it is, the number of traffic fatalities pales in comparison to the number of people injured. In 2010, 3,341,640 people (377,798 of them pedestrians or cyclists) suffered non-fatal traffic-related injuries,[356] almost 100 times the number that died.[357]

Several studies have investigated the relationship between vehicle crashes and the form of our built environment. A review of the literature shows that as population density increases, so do crash rates per VMT. However, increased population density is also associated with reductions in per capita VMT (reducing per capita crash rates) and reductions in travel speed (which is the primary determinant of crash severity).[358] One study considered the association between crash fatalities and a sprawl index[359] across 448 counties in the 101 largest metropolitan areas.[360] Places with high-density residential

---

[354] Subramanian 2011

[355] Subramanian 2011

[356] Centers for Disease Control National Center for Injury Prevention and Control, *WISQARS Nonfatal Injury Reports* n.d.

[357] Centers for Disease Control National Center for Injury Prevention and Control, *WISQARS Fatal Injury Reports* n.d.

[358] Ewing and Dumbaugh 2009

[359] Many studies reported in this document use one of several composite sprawl indices. They incorporate multiple aspects of urban form such as density, land use mix, connectivity, and centeredness, in an attempt to develop a concise quantitative description of how sprawling or compact an area is.

[360] Ewing, Schieber, and Zegeer 2003

development; homes, shops, and workplaces in the same area; a distinct downtown or activity center; small blocks; and well-connected streets scored highest on the index. The study found that it is safer to be in places with these characteristics. Every 1 percent increase in the index was associated with a 1.5 percent decrease in traffic fatalities. The decrease was greater, 1.5 to 3.6 percent, if only those fatalities involving pedestrians were included, after adjusting for the amount of time people walk in different locations.

Poor and minority populations suffer a disproportionate number of pedestrian fatalities. The causes of the disparity are complex, possibly involving differences in the amount of walking based on access to personal vehicles and use of public transportation, street design in areas where poor and minority individuals live, and cultural factors such as experience with traffic.[361]

3.8   Summary

As the U.S. population has grown, we have developed land that serves important ecological functions at a significant cost to the environment. Development has destroyed, degraded, and fragmented habitat. Water quality has declined. Air quality in many areas of the country is still adversely affecting human health. The heat island effect and global climate change illustrate just how complex and far-reaching the impacts of our built environment are. Community design can make it difficult for people to get adequate physical activity, engage with neighbors, and participate in community events. It can also increase the risk of injury or death from a vehicle crash.

The next chapter discusses the impacts of different development patterns to help communities understand how their choices about where and how to develop can better protect the environment and residents' health and safety.

---

[361] Frumkin 2006

Chapter 4. Effects of Different Types of Development on the Environment

In over 40 years since the creation of EPA, the country has made significant advancements in cleaning our air and water and protecting critical lands and habitat. However, environmental problems remain, and the built environment affects our ability to address these problems. Where and how we build affect attainment of national environmental goals in each of the following areas:

- **Taking action on climate change and improving air quality**—Our ability to reduce air pollution emissions and greenhouse gases depends in part on how much people drive and the energy efficiency of our buildings.

- **Protecting America's waters**—Our ability to protect water quality depends in part on the amount of impervious surface we create in our developed areas and how effectively we can preserve undeveloped land.

- **Cleaning up communities and advancing sustainable development**—Our ability to create healthy and safe communities depends in part on whether we are able to clean up and repurpose developed areas that have been neglected.

- **Ensuring the safety of chemicals and preventing pollution**—Our ability to prevent toxic chemicals from entering the environment depends in part on the materials we use to build our homes and businesses.

- **Enforcing environmental laws**—Environmental laws help us achieve all of the above, particularly by ensuring the protection of low-income, minority, and tribal communities that are disproportionately affected by pollution.[362]

EPA's environmental goals and national efforts have historically focused on reducing tailpipe and smokestack emissions. As Congress passed legislation to address environmental and human health threats, EPA responded with regulations aimed largely at controlling the most obvious risks, such as pollution from large industries. It has become increasingly clear that this approach alone is not sufficient. EPA now recognizes that "development and building construction practices may result in a broad range of impacts on human health and the environment" and promotes sustainable communities that "balance their economic and natural assets so that the diverse needs of local residents can be met now and in the future with limited environmental impacts."[363]

As communities consider how to grow, they are looking for strategies that will protect the environment while encouraging new economic opportunities and improving quality of life. Different development patterns can affect the environment in different ways. The direct effects of habitat consumption and disruption are well documented and widely accepted. By contrast, the effects of different development patterns on travel and emissions are somewhat less understood, and the exact magnitude of these

---

[362] EPA, *FY 2011-2015 EPA Strategic Plan* 2010
[363] EPA, *FY 2011-2015 EPA Strategic Plan* 2010, pp. 15-16

effects is still subject to some debate. This section synthesizes research findings, noting where greater or less certainty exists about linkages between causes and effects and the relative magnitude of effects. Where previous sections described broad land use and transportation trends and impacts, this section focuses on impacts of conventional development versus more environmentally sensitive development.

Certain characteristics of the built environment are associated with beneficial environmental results. They can be divided broadly into two categories:

- **Where we build** involves locating development in a region or land area. It includes safeguarding sensitive areas such as riparian buffers, wetlands, and critical habitat from development pressures; directing new development to infill, brownfield, and greyfield sites[364] to take advantage of existing infrastructure and preserve green space; and putting homes, workplaces, and services close to each other in convenient, accessible locations.
- **How we build** includes developing more compactly to preserve open spaces and water quality; mixing uses to reduce travel distances; designing communities and streets to promote walking and biking; and improving building design, construction, and materials selection to use natural resources more efficiently and improve buildings' environmental performance.

These elements are interrelated and often work most effectively in combination with each other rather than individually. For example, encouraging compactness through infill or brownfields redevelopment often facilitates mixed-use development and creates the type of environment that makes transit use, walking, and bicycling easier and more appealing. Developing more compact development can help safeguard environmentally sensitive areas.

Incorporating one beneficial element without others could have minimal effects or possibly prove detrimental to environmental goals. For example, increasing density without protecting environmentally sensitive areas or improving transit access could result in increased water quality impacts, traffic congestion, or air quality problems. Likewise, locating energy-efficient homes far from an employment center and other amenities can increase total energy expenditures as residents drive longer distances to work and other activities.

Because most practices work synergistically with one another, isolating the effects of one from another can be difficult. In general, the studies presented in this chapter were chosen because they attempt to isolate the effects of individual strategies. However, studying one technique in isolation is often nearly impossible, and the value of doing so is somewhat limited, as most practices are used in combination with others. Nonetheless, although findings might differ on the magnitude of the effects of different practices, the evidence is overwhelming that some types of development yield better environmental results than others. Researchers have estimated that as much as two-thirds of the development that will

---

[364] Greyfields are formerly economically viable sites that suffer from disinvestment, often retail shopping malls, strip centers, or office parks. By definition, greyfields have no environmental contamination (or they would be considered brownfields), and they can offer important attributes for developers such as a large lot size, existing infrastructure, and an accessible location.

exist in 40 to 45 years does not exist today,[365] meaning that decisions we make about how and where that development occurs could significantly affect our health and the health of the environment.

4.1   Where We Build

Where we locate development has significant impacts on environmental resources. Development in and adjacent to already-developed areas can help protect natural resources like wetlands, streams, coastlines, and critical habitat. Protecting natural lands from development can reduce impacts on habitat and water resources, reduce VMT and associated air pollutants, foster social connections, and improve human health.

This section covers the following aspects of where we build:

- Safeguarding sensitive areas.
- Infill development in built-up areas.
- Focusing development around existing transit stations.

4.1.1   Importance of Safeguarding Sensitive Areas

Using land for development more efficiently makes it easier to conserve natural lands. Communities can encourage development in places where transportation, utilities, and public services such as schools and hospitals already exist; in areas adjacent to built-out communities; and on brownfields and greyfields. Encouraging development in these places, where infrastructure already exists and the environment has already been disturbed, can reduce pressure to develop farmland and sensitive natural areas such as wetlands, flood plains, mature forests, drinking water source areas, and shorelines. These areas serve important ecological functions (see Section 3.1), and their preservation provides valuable scenic and recreation areas.

In addition, development in sensitive areas like coastal zones can exacerbate the ecological effects of sea level rise and other impacts of global warming.[366] Climate change and habitat destruction from development and shoreline stabilization are some of the most serious threats to coastal species and ecosystems.[367] Protecting coastal areas from development can also help keep people away from areas prone to natural hazards and limit the demand for shoreline stabilization.[368] As global warming increases, hurricanes, flooding, and wildfires are likely to increase,[369] so avoiding development in areas most vulnerable to these natural disasters such as coastal areas, flood plains, and wildfire-prone forests will help keep property and people out of harm's way. Communities will also need to consider that areas

---

[365] Ewing, Bartholomew, et al. 2008
[366] Defeo, et al. 2009
[367] Crain, et al. 2009. Other threats that consistently rank highly are invasive species and bottom fishing.
[368] Defeo, et al. 2009
[369] National Research Council of the National Academies 2011

Exhibit 4-1: A mature bottomland hardwood forest filled with native plants. The Blockston Branch of the Tuckahoe Creek at the Atkins Arboretum in Ridgeley, Maryland, provides a wealth of ecosystem services, including water and air purification and habitat for quail, bluebirds, foxes, and turkeys.
Photo source: EPA

that are currently considered safe from flooding, storms, or wildfires could become vulnerable as the climate changes.

Communities have used a variety of approaches to safeguard farmland, forests, coastal zones, and other types of open space.[370] In the most direct approach, local, regional, state, and federal agencies acquire parks, recreation areas, forests, wildlife refuges, and wilderness areas to ensure their long-term protection and maintenance. Limited resources and political considerations preclude most governments from owning all critical land areas they might want to preserve. Often communities supplement these efforts with regulatory approaches that discourage or limit the amount and type of development that can occur in sensitive areas. For example, many communities use zoning to direct development to certain areas and to establish allowable development densities. Local or regional comprehensive plans can also encourage development in certain places; for example, by designating which areas are eligible for public infrastructure. In addition, many communities complement direct ownership and regulation with incentives to promote private stewardship of important natural resources. Tax policies can provide financial incentives that encourage or discourage development in certain areas. Streamlined permitting and review processes for projects in already developed areas can also encourage development in those locations.

A review of how effectively these types of public policies protect open space found that while robust analyses of different policies are generally lacking, communities that have been most successful in safeguarding their sensitive lands have multiple, complementary policies aimed at achieving this goal. Coordination across different levels of government and jurisdictions at a regional scale is also critical for success. Finally, the review also found that meaningful involvement of residents and other stakeholders

---

[370] Bengston, Fletcher, and Nelson 2004

in developing a community vision and the tools to implement it is an essential component of effective land preservation strategies.[371]

4.1.2 Importance of Infill Development

Infill development occurs in locations where some development has already taken place and infrastructure is already available. It includes redeveloping old buildings and facilities and putting undeveloped space in developed areas into productive use. Infill development reduces development pressure on outlying areas, helping to achieve the goal of safeguarding critical lands. When infill development occurs near existing transit infrastructure or near employment centers and other destinations, it can also help slow growth of or even reduce VMT by reducing the amount people need to drive.[372] In fact, development's location relative to the surrounding places people want to go is one of the most important determinants of how much people travel by car (see Section 4.2.5).

Exhibit 4-2: Infill development. Developers of the Matrix Condos in Washington, D.C., converted a long-abandoned former auto dealership into a mixed-use project by building on adjacent empty lots and adding two stories to the existing building, set back from its façade. The condo residents are within walking distance of downtown, a grocery store, multiple restaurants, and other local businesses. A bus stop is right outside their front door.
Photo source: EPA

Infill Potential

The amount of land available for infill is difficult to quantify. Many cities in the United States have populations far from their peak, suggesting that existing infrastructure and space in these places could accommodate more people. The number of abandoned lots and buildings only underscores this observation. A 2009 survey of mayors from 53 cities estimated the number of vacant and abandoned properties. The highest estimates were 15,078 in Las Vegas; 13,500 in St. Louis; 8,306 in Louisville, Kentucky; 7,700 in Port St. Lucie, Florida; and 7,000 in Cape Coral, Florida. The lowest estimates were 25 in Stow, Ohio; 21 in Menlo Park, California; and 11 in Bell Gardens, California.[373]

The U.S. Census collects information on the number of vacant housing units (Exhibit 4-3). At 7.9 percent, Connecticut has the lowest rate of vacancy, while both Maine and Vermont have the highest rates of more than 20 percent, likely due to a high percentage of seasonal, recreational, or occasional use homes.

---

[371] Bengston, Fletcher, and Nelson 2004
[372] Landis, et al. 2006
[373] United States Conference of Mayors 2009

State	Total Housing Units in 2010	Percent Vacant	State	Total Housing Units in 2010	Percent Vacant
Maine	721,830	22.8	South Dakota	363,438	11.3
Vermont	322,539	20.5	Tennessee	2,812,133	11.3
Florida	8,989,580	17.5	Minnesota	2,347,201	11.1
Arizona	2,844,526	16.3	Colorado	2,212,898	10.8
Alaska	306,967	15.9	Kentucky	1,927,164	10.8
Delaware	405,885	15.7	Rhode Island	463,388	10.7
South Carolina	2,137,683	15.7	Texas	9,977,436	10.6
New Hampshire	614,754	15.6	Indiana	2,795,541	10.5
Montana	482,825	15.2	Utah	979,709	10.4
Michigan	4,532,233	14.6	Ohio	5,127,508	10.2
Nevada	1,173,814	14.3	District of Columbia	296,719	10.1
North Carolina	4,327,528	13.5	Pennsylvania	5,567,315	9.9
West Virginia	881,917	13.4	Kansas	1,233,215	9.8
Wyoming	261,868	13.4	New York	8,108,103	9.7
Alabama	2,171,853	13.3	Nebraska	796,793	9.5
Idaho	667,796	13.2	New Jersey	3,553,562	9.5
Wisconsin	2,624,358	13.1	Maryland	2,378,814	9.3
Arkansas	1,316,299	12.9	Massachusetts	2,808,254	9.3
Mississippi	1,274,719	12.5	Oregon	1,675,562	9.3
Hawaii	519,508	12.4	Virginia	3,364,939	9.2
Missouri	2,712,729	12.4	Washington	2,885,677	9.2
Georgia	4,088,801	12.3	Illinois	5,296,715	8.7
Oklahoma	1,664,378	12.3	Iowa	1,336,417	8.6
New Mexico	901,388	12.2	California	13,680,081	8.1
Louisiana	1,964,981	12.0	Connecticut	1,487,891	7.9
North Dakota	317,498	11.4	**TOTAL**	**131,704,730**	**11.4**

Exhibit 4-3: Percentage of housing units vacant, 2010.
Source: U.S. Census Bureau *Housing Characteristics: 2010* 2011

A 2006 analysis estimated that California had nearly half a million potential infill parcels, defined as vacant or underused lots within existing city boundaries or in unincorporated places with at least 2.4 residential units per acre.[374] Together, these parcels occupied 220,000 acres. Maintaining existing residential densities in neighborhoods, these sites could accommodate about 1.5 million additional housing units, 25 percent of the state's projected need over 20 years. If residential densities increased to levels compatible with existing residential character (depending on location) and higher-density housing was built near transit, the researchers estimated that an additional 4 million housing units could be accommodated on existing infill sites. This scenario would meet 100 percent of the state's projected need over 20 years and avoid development on 350,000 acres of undeveloped land.

An EPA analysis of residential construction trends supports the idea that many metropolitan regions can support a significant amount of infill development.[375] Among 20 metropolitan regions, 21 percent of new home construction between 2000 and 2009 occurred in previously developed areas. In some regions, the figure exceeded 60 percent (Exhibit 4-4).

---

[374] Landis, et al. 2006
[375] EPA, *Residential Construction Trends in America's Metropolitan Regions* 2012

Exhibit 4-4: Percentage of new home construction that is infill, 2000-2009.
Image source: EPA, *Residential Construction Trends in America's Metropolitan Regions* 2012

Certain types of underdeveloped land, greyfields, brownfields, and hazardous waste sites, have received significant attention as both problems and opportunities because their redevelopment has the potential to provide multiple community benefits. As noted in Section 3.2, reliable national area data for these types of sites are not available.

Environmental Benefits of Cleaning up Brownfields and Hazardous Waste Sites
The cleanup and redevelopment of brownfields and hazardous waste sites can bring substantial environmental and health benefits. The most obvious environmental benefit of land cleanup is the safe disposal of environmental contaminants. Land cleanup activities also provide the opportunity to reuse the land for new development, removing pressure to develop in natural areas that have important ecological functions. Land reuse prevents many of the greenhouse gas emissions that would come from the construction of new transportation and other infrastructure necessary to serve the development. Land cleanup can also provide opportunities to reuse and recycle construction and demolition debris from buildings in ways that protect human health and the environment. Reusing materials and land avoids the need to devote any additional land to waste disposal. Finally, certain site remediation techniques can provide additional environmental benefits. For example, adding organic matter to soil increases carbon dioxide sequestration, enhances vegetation growth, and can even make productive use

of what might otherwise be a waste product.[376] Greyfield, brownfield, and hazardous site redevelopment can dramatically improve environmental quality and community life, since undeveloped sites can be health threats or discourage further investment in the area. Investments in site assessment and cleanup can encourage additional investment in cleanup and redevelopment nearby, further improving environmental outcomes.

4.1.3 Benefits of Focusing Development Around Transit Stations

Transit systems with fast, frequent, and dependable service attract riders and can reduce VMT, reducing air pollutant and greenhouse gas emissions. A bus carrying 20 passengers consumes only about one-third of the energy that would be needed if each passenger drove a private vehicle.[377] Transit also provides mobility to people who cannot or choose not to drive, which can help low-income people find and keep jobs and helps older residents maintain their independence when they can no longer drive.

Transit-oriented development (TOD) locates housing, shopping, and employment near transit stations. It makes transit a more convenient and practical form of transportation and can be a catalyst for other land use changes that benefit the environment. A review of the literature found that under certain conditions, light rail could increase the density of development. The best conditions for spurring development occur when light rail is constructed in areas that are growing, when it significantly improves access to the areas it serves, when there is developable land around stations, and when local land use policies support TOD.[378]

Several studies have attempted to quantify the impacts of TOD on transit use. For example, a study of 26 TOD projects in California found that people living within a half-mile of a rail station were about four times as likely to commute by rail as those living between one-half and 3 miles away and six times as likely to commute by rail as those living more than 3 miles from the station. Residents who are inclined to use transit often choose to live near a transit station, which explains part of the difference. However, other important factors included whether residents were able to access jobs easily by highways (which decreases transit use) and whether there was a walkable street grid at places residents can reach by transit (which increases transit use).[379] In addition to the location of housing, the location of employment is also important. Employees who work at locations near rail stations were three times as likely to commute by public transportation as those working at places without rail access. Frequent bus service from stations to office sites, employer subsidies for transit costs, and scarcity of parking were also important factors in increasing transit use.[380]

---

[376] EPA, *Opportunities to Reduce Greenhouse Gas Emissions through Materials and Land Management Practices* 2009

[377] A car has an energy intensity of 5,342 Btu per vehicle mile. Twenty drivers would therefore use 106,840 Btu per mile compared to 35,953 Btu per mile if all rode together on a bus (Davis, Diegel, and Boundy 2012, Table 2.12).

[378] Handy 2005

[379] Cervero 2007

[380] Cervero 2006

Portland, Oregon, has used TOD as one of many policy tools to encourage transit use (Exhibit 4-5). A study of commuting behavior in census block groups in Portland found that commuters living near light rail or bus service in areas with a higher proportion of mixed use were more likely to use public transit, walk, or bike to work. Commuters closer to freeway interchanges in areas with a higher share of land used for single-family homes were more likely to drive alone. However, the study found no correlation between commuting behavior and TOD, although it measured the degree to which an area's development was transit oriented based only on whether it was within one mile of a light rail station and either housing or job density.[381] A study based on resident surveys in Portland showed that those living in TODs were more than 2.5 times as likely to commute by transit as other Portland residents. Nearly 20 percent of transit commuters living in TODs had switched from other modes of travel, while just 4 percent of commuters living in TODs switched from transit to another mode of travel.[382]

Exhibit 4-5: Mixed-use, multifamily building.
Developments on the streetcar line in the Pearl District of Portland, Oregon, give residents easy access to universities, a major hospital, and the central business and shopping districts.
Photo source: Kyle Gradinger via flickr.com

A more comprehensive evaluation of TOD across cities looked at 17 TOD projects of different sizes in four areas of the country: Philadelphia and northeast New Jersey; Portland, Oregon; metropolitan Washington D.C.; and the San Francisco Bay Area's East Bay. Researchers compared actual vehicle trip rates with those predicted using the Institute of Transportation Engineers manual, which is often used to project local traffic and parking impacts and set impact fees for development projects. Across the 17 TOD projects, about 47 percent fewer trips occurred than projected. The largest reductions in vehicle travel occurred at TOD projects closest to central business districts with the highest residential densities.[383]

A 2008 review of the literature found that people living in TODs are two to five times more likely to use transit for commuting and nonwork trips than other people living in the same region. Many of the transit users at TOD sites previously used transit and presumably chose to live in a TOD for its transit accessibility. However, TOD increased transit use by up to 50 percent for people who had no prior

---

[381] Jun 2008
[382] Dill 2008
[383] Arrington and Cervero 2008

transit access. Transit use varies considerably across regions, determined primarily by how well transit links destinations.[384]

In addition to the location of development relative to transit, the location of transit itself is important. The effects of transit on overall development patterns depend on where it is located and what development patterns it serves.[385] For example, transit can make it convenient for people to live far from where they work and lead to new development in previously undeveloped land. Building public transit that serves already-developed areas can help attract growth to areas where it can help mitigate overall environmental impacts from development rather than create new ones.

4.2   How We Build

How we build also influences development's impact on the environment and human health. The way we build our cities and towns, including the details—the scale and relationship of buildings, blocks, streets, and public places—determines how close homes are to workplaces, how many destinations can be reached comfortably and pleasantly by walking or biking, and whether frequent bus or rail transit is practical. The design and connectivity of transportation networks can either make it easy to get around by walking, biking, transit, and short driving trips, or require people to drive for every trip. Green design—of neighborhoods, streets, and buildings—can help clean and manage stormwater, use energy and water more efficiently, and improve air quality.

The relationship between the built environment and travel behavior is complex. An extensive body of literature attempts to measure the impacts of the built environment on travel behavior, producing not just multiple reviews but also reviews of the many reviews.[386,387] Measuring the effects of various land use characteristics on VMT is challenging because these characteristics change slowly and have delayed effects. It is also difficult to identify treatment and control groups for rigorous experimental design.[388] In addition, differences in methodology, units of analysis, scale of study, and the characteristics considered make it difficult to draw overall conclusions.[389]

In spite of indications that many individual components of the built environment have relatively small effects on travel behavior,[390] small effects of individual variables can be cumulative.[391] These effects tend to have local impacts at first[392] that increase over time.[393] Many of the environmental challenges we face today require a comprehensive approach in which strategies with fairly modest results can

---

[384] Arrington and Cervero 2008

[385] Handy 2005

[386] Gebel, Bauman, and Petticrew 2007

[387] Ewing and Cervero 2010

[388] Salon, et al. 2012

[389] Bhat and Guo 2007

[390] Salon, et al. 2012

[391] Kuzmyak, et al. 2003

[392] Kuzmyak, et al. 2003

[393] Dulal, Brodnig, and Onoriose 2011

cumulatively play important roles in protecting our environmental resources and ensuring the needs of future generations can be met.

When interpreting research on the built environment, one important factor is the extent to which people choose to live in neighborhoods based on their travel options. Results that appear to suggest that the built environment could change people's inclination to use transit, bike, or walk might instead mean that the built environment merely makes it possible to use these transportation options.[394] Researchers have called the tendency of people to live in places that accommodate their travel preferences *self selection*. If self selection alone can explain an association between the built environment and travel behavior, then land use changes are likely to change travel behavior only to the extent of the unmet demand for neighborhoods that can accommodate different forms of travel—and

research has demonstrated that indeed unmet demand for more pedestrian-oriented neighborhoods exists.[395,396] Many studies have documented self-selection effects and highlight the importance of controlling for this factor when evaluating associations between the built environment and travel behavior,[397,398] although some studies have found that its impacts are modest or non-existent.[399,400] There is some evidence that elements of the built environment that improve walkability and transit accessibility actually have *smaller* impacts on the travel behavior of people who seek walkable, transit-accessible neighborhoods than on the travel behavior of people who do not prioritize

Exhibit 4-6: The Castro District, San Francisco. The Castro District is one of the city's most vibrant and cohesive neighborhoods. Small and medium-sized single-family Victorian homes and small apartment buildings create a compact neighborhood where residents can walk to shops, businesses, parks, and movie theaters.
Photo source: EPA

[394] Handy 2005

[395] Levine, Inam, and Torng 2005

[396] Levine and Frank 2007

[397] Cao, Mokhtarian, and Handy 2009

[398] Bhat and Guo 2007

[399] Chatman 2009

[400] Ewing and Cervero 2010

these attributes because people who prefer to walk or take transit tend to do so regardless of their neighborhood characteristics.[401]

Describing discrete components of the built environment can be challenging because factors overlap and variables correlate with one another. Thus, many studies have similar findings, although the variables they investigate vary. The following sections cover several attributes of the built environment that are relatively well studied across the United States:[402]

- Compact development.
- Mixed-use development.
- Street connectivity.
- Community design.
- Destination accessibility.
- Transit accessibility.
- Green building.

4.2.1 Compact Development

Compact development generally means using less land area to satisfy the needs of a population. The more compact a community, the less land is needed for development, reducing environmental impacts. In more compact areas, people can travel shorter distances for everyday activities, and it is easier to walk or bike to those destinations. In addition, compact development makes public transit, sidewalks, and bike paths more practical and cost-effective because destinations are closer together.

Many studies have tried to investigate the effects of compact development using population density as a metric. However, higher population density rarely occurs without at least some other confounding factors, such as a mix of land uses, access to public transportation, and the

Exhibit 4-7: Compact residential neighborhood. Many city streets in Philadelphia are lined with single-family rowhouses and small apartment buildings. Narrow streets and shade trees make walking to nearby commercial areas convenient and pleasant.
Photo source: EPA

---

[401] Chatman 2009

[402] This literature review excluded research based solely on data from locations outside of the United States, except when geographic location is unlikely to affect the results (e.g., for performance of building energy efficiency technologies).

presence of sidewalks and other characteristics that would likely influence how people choose to travel. Indeed, density is often a prerequisite for community characteristics such as better transit service or a higher diversity of land uses.[403] Public transit is more cost-effective the more people and destinations it can serve in a given area. In addition, more people in a given area can support more types of businesses and institutions. This section looks at studies that have attempted to isolate the effects of compact development on the environment.

Vehicle Miles Traveled

The density of people in a community—i.e., the number of people that live per unit of land—is a good indicator of how much people will drive and how much they will get around by foot or by bike. Density is strongly correlated with many of the characteristics that make driving long distances unnecessary and make other forms of transportation appealing alternatives to driving: mixed residential, commercial, and institutional areas; destinations situated close together; and sidewalks.[404] In general, the greater the population density of an area, the less the area's residents tend to drive.[405]

The majority of the research on compact development's effect on the environment looks at the issue by considering its influence on VMT and the negative environmental and human health effects associated with vehicle travel. Reviews of the literature have noted for decades the association between density and vehicle travel, measured by both the number of per capita trips and VMT.[406] As expected, along with the decline in VMT, higher densities are associated with increases in other forms of travel, particularly transit and walking.

The National Research Council considered the issue in a 2009 report.[407] Based on a review of the literature, this report confirmed that higher residential and employment density result in less driving and more transit use and walking. After accounting for socioeconomic factors and adjusting for self selection, the report concluded that doubling residential density across a metropolitan region could reduce household VMT by between 5 and 12 percent. Reductions of as much as 25 percent would be possible if increased residential density occurred with other changes such as increased employment density, improved public transit service, a mix of land uses, and other changes that encourage alternatives to driving such as parking demand management and street design that encourages walking and biking. The report noted that the air quality benefits from reduced VMT would occur in addition to other environmental benefits from more compact development, such as more land conservation and reductions in stormwater runoff.

Other, more recent studies and reviews have confirmed the general pattern that increased density reduces VMT. For example, a 2012 meta-analysis of studies on the relationship between density and travel behavior in the United States and Europe confirmed the decline in VMT with higher density holds

---

[403] Kuzmyak, et al. 2003

[404] Kuzmyak, et al. 2003

[405] Transportation Research Board of the National Academies 2003

[406] Kuzmyak, et al. 2003

[407] National Research Council of the National Academies, *Driving and the Built Environment* 2009

for both regions, although it is considerably stronger in Europe.[408] However, the size of the effect is still uncertain. One study analyzed the relationship between VMT and both population and employment density across 370 urbanized areas in the United States.[409] A 10 percent increase in population density was associated with a 3.8 percent decrease in VMT per capita, while employment density had more modest effects. A modeling study considered how VMT would change in response to changes in development patterns, incorporating information about where people choose to live to correct for self selection. The model results show that increasing residential density, jobs per capita, and per capita expenditures on public transit operations each by 10 percent could reduce annual VMT from 22,182 miles to 17,782 per household, a reduction of about 20 percent.[410]

The most recent reviews note that high residential and employment density are most strongly associated with reduced VMT when coupled with other factors such as mixed land uses and public transit service.[411] Such factors that tend to co-occur with density appear to be of considerable importance. Density by itself generally shows only relatively modest effects. A meta-analysis[412] of travel literature that isolated individual measures of the built environment found that after controlling for all other measures, population and job density actually had the weakest association to travel behavior among those that are significant.[413] A minimum population density is necessary for many of the other factors that influence travel behavior (discussed in the following sections), but alone, it is likely not sufficient.

Exhibit 4-8: Heavy traffic. Interstate 94 in Minneapolis carries commuters to and from downtown, contributing air pollution to the entire region.
Source: drouu via stock.xchng

Vehicle Miles Traveled and Air Pollution
Several studies have considered the link between density, VMT, and air pollution. For example, a study of 45 of the largest metropolitan regions over 13 years found a positive correlation between a quantitative sprawl index and both emissions of ozone precursors and the number of high ozone

---

[408] Gim 2012
[409] Cervero and Murakami 2010
[410] Chattopadhyay and Taylor 2012
[411] Dulal, Brodnig, and Onoriose 2011
[412] Meta-analyses use summary statistics from individual primary studies as the data points in a new analysis. They help establish overall relationships when the results of the individual studies were inconsistent or their study designs, statistical techniques, and/or study populations varied (Gim 2012).
[413] Ewing and Cervero 2010

days.[414] The study found that the least compact regions (those scoring highest on the sprawl index[415]) had 60 percent more high ozone days than the most compact regions. Of the variables included in the sprawl index, density was most strongly associated with this effect: each standard deviation increase in density was associated with an average of about 16 fewer high ozone days per year. The study found that the positive correlation between the sprawl index and the number of high ozone days held even when controlling for annual emissions of ozone precursors. Since transportation is a significant contributor to ozone precursor emissions, the study suggests that lower VMT in more compact metropolitan regions cannot entirely explain the lower levels of ozone air pollution.

Another study estimated the carbon dioxide emissions of 100 of the largest metropolitan regions of the United States.[416] Per capita emissions varied considerably across regions, with the highest emitting region (Bakersfield, California) emitting almost 2.5 times as much carbon dioxide per resident as the lowest-emitting region (New York, New York/New Jersey). Differences were even greater when emissions were calculated based on gross metropolitan product, the metropolitan region's equivalent of gross domestic product. Emissions per gross metropolitan product in the highest-emitting region (Riverside, California) were almost five times the emissions in the lowest-emitting region (Bridgeport, Connecticut). Analysis of differences showed that the amount of trucking activity in a metropolitan region could explain much of this variation. However, overall population and employment density, as well as how people and jobs were distributed across the region, also helped explain differences among regions. A higher density of people and jobs per acre was associated with lower carbon dioxide emissions.

Development that is more compact can reduce greenhouse gas and other air pollution emissions, not just by reducing travel, but also by reducing the amount of infrastructure needed. Smaller, more compact buildings use less energy for heating and cooling. A study compared the greenhouse gas emissions associated with a high-density residential area in downtown Toronto with a low-density suburb. The study included emissions associated with transportation, infrastructure construction, and building operations. It found that the low-density neighborhood was between two and 2.5 times more energy intensive per capita than the high-density neighborhood.[417]

A 2008 paper reviewed four different types of literature on the relationship between development patterns and travel behavior: studies that aggregated travel behavior at a regional level, studies that looked at individual travel behavior, regional simulations, and project-level simulations.[418] When enough data permitted, the authors conducted a meta-analysis of the results. The results suggested that within a region, compact development in combination with other strategies—such as increasing land use mix and ensuring that destinations are easily accessible from around the region—could reduce VMT

---

[414] Stone 2008

[415] According to the modified sprawl index used in this study, higher scores denote higher levels of sprawl, unlike other studies discussed earlier in which higher scores denote lower levels of sprawl.

[416] Southworth and Sonnenberg 2011

[417] Norman, MacLean, and Kennedy 2006

[418] Ewing, Bartholomew, et al. 2008

significantly, by as much as 20 to 40 percent. Nationally, total VMT could be reduced by 10 to 14 percent, leading to a 7 to 10 percent decline in total U.S. transportation carbon dioxide emissions.

Regional Versus Neighborhood Air Quality

In addition to regional air quality, communities are concerned about neighborhood-level air quality. Some studies have looked within regions at population-weighted air pollution exposures, which represent the average exposure level of the area's population.[419]

The relationship between the location of pollution sources and air quality impacts is complex. Air pollutants have different sources, concentration patterns around sources, and sensitivity to climate and weather conditions. *Primary pollutants* are emitted directly to the atmosphere, while *secondary pollutants* result from chemical reactions in the atmosphere. Concentrations of some pollutants are highest near the source and tend to drop as distance from the source increases. Pollutants with this pattern include primary pollutants, such as carbon monoxide, and those secondary pollutants that form quickly, such as nitrogen dioxide. Concentrations of other pollutants tend to have a more even distribution because they are transported downwind over wide areas. Pollutants with this pattern include those secondary pollutants such as ozone that can take hours to form as their precursors move downwind and undergo chemical reactions.[420] As a result, local concentrations of this type of pollutant might not be well-correlated with local emissions of its precursors. For example, the highest ozone concentrations are often downwind of areas where VOC and NO_x emissions are highest.[421] Understanding these differences is critical to evaluating studies examining the relationship between compact development and the concentrations of different air pollutants.

One study investigated the relationship between compact development and regional and neighborhood levels of ozone and $PM_{2.5}$ in 80 U.S. metropolitan regions.[422] It found that regional ozone concentrations were lower in more compact areas (as measured by a sprawl index), but population-weighted ozone exposures were higher because people tended to be concentrated around air quality monitors recording higher pollution levels. Although the average regional concentration of $PM_{2.5}$ was not correlated with the sprawl index, population-weighted exposures to $PM_{2.5}$ were higher in compact regions, as was the case for ozone. The study also found that neighborhoods with higher percentages of minority and poor residents had higher concentrations of both ozone and $PM_{2.5}$. This study did not examine how population influenced pollution levels, so it is unclear from this study whether differences in vehicle activity, the timing and location of emissions, the distance between emissions and people, or other factors might explain the results.

---

[419] Population-weighted air pollution exposures are determined by multiplying the population in the "neighborhood" around each air quality monitor (e.g., the population living within a half mile of a monitor) by the monitor's reading for every monitor in an area, adding those results, and then dividing by the total population of the area.

[420] Marshall, McKone, et al. 2005

[421] Other complex interactions between the precursors of ozone and secondary PM components influence the interpretation of these studies. For example, in many urban centers, emissions of NO_x reduce concentrations of ozone within the city, but raise concentrations in locations downwind.

[422] Schweitzer and Zhou 2010

Another study of 111 U.S. urban areas found that population-weighted exposure to $PM_{2.5}$ and aggregate pollutant levels were higher in areas with higher population density.[423] However, population-weighted ozone exposure was not associated with population density in this study, and population centrality (see Section 4.2.5), another common attribute of compact regions, was associated with lower population-weighted exposure to $PM_{2.5}$ and ozone. Thus, while development patterns can clearly influence air pollution, compact development encompasses several attributes, and density alone might not be its best measure.

There are several considerations to take into account when reviewing the results of these and other studies. First, it is important to consider the quality and purpose of the air quality monitoring data that is the basis of findings for neighborhood population exposure. In the study of 111 urban areas, the air quality data used for ozone and $PM_{2.5}$ were from 1990 and 2000, respectively.[424] Since those data were collected, many of the areas in this study have achieved significant emission reductions despite increasing population. Furthermore, the majority of the vehicles on the road in 1990 have been replaced with vehicles meeting more stringent emission standards. More recent data could lead to different results and conclusions because many air pollutant effects are non-linear. In addition, the air quality monitoring networks in the United States have been designed to meet statutory and regulatory objectives[425] rather than to support this type of research. In fact, the study on the relationship between compact development and regional and neighborhood levels of ozone and $PM_{2.5}$ in 80 U.S. metropolitan regions discusses several methods of sampling and grouping existing data to address such limitations.[426] Finally, historical development patterns mean that residential neighborhoods are often located near ports, heavy industry, and other major emissions sources. The relationship between population-weighted pollution exposures and the compactness of a region reflects in part this history and does not imply causality. All of these factors complicate the interpretation of results of these studies.

One useful conclusion from this body of research is that better air quality outcomes could be achieved if communities coupled compact development with careful consideration of that development's location within a region. Locating new, compact development away from major sources of emissions, such as freight corridors or power plants, would help minimize exposure to these pollution sources while reducing pollution regionally. For pollutants where air quality concentrations decrease with distance from their source, relatively small adjustments in the location or design of development could have large effects on air pollution exposures. When advocating for more compact development to help reduce VMT, preserve habitat, and increase physical activity, planners could take differences in neighborhood air quality into account, where data are available to evaluate potential locations or better site design.

In addition to considering the location of new compact development relative to emissions sources, other strategies can help avoid high neighborhood-level concentrations. For example, a higher proportion of

---

[423] Clark, Millet, and Marshall 2011
[424] Clark, Millet, and Marshall 2011
[425] EPA, *The Ambient Air Monitoring Program* n.d.
[426] Schweitzer and Zhou 2010

lower-emitting vehicles in city centers and use of indoor air filtration systems reduce individual air pollution exposures.[427] Strengthening the impact of density on VMT through complementary urban design features could also help decrease both regional pollution levels and per capita exposure.[428] Sections 4.2.2 through 4.2.6 discuss the types of strategies that research has shown to strengthen this relationship. Improving air quality for the largest number of people would likely require additional emissions reductions from vehicles and other sources as well as supporting denser, mixed-use districts where walking, biking, and transit use reduce the need to drive.[429]

Heat Island Effect

As discussed in Section 3.5, the heat island effect is associated with the built environment. Locating tall buildings near one another can contribute to the heat island effect by limiting wind circulation and

creating large areas of impermeable surfaces that absorb the sun's heat. However, in other ways, building compactly to minimize developed area could potentially lessen the heat island effect. Compact development reduces impervious area that can absorb the sun's heat and prevent evaporative cooling. It does this in several ways, including by using taller buildings that require less roof per amount of living space, smaller lots that require fewer miles of roads and sidewalks, and homes nearer to the street to reduce driveway lengths.

One study of the Atlanta region looked at how single-family residential density affected an indicator of surface heat island formation.[430] The results showed, somewhat counterintuitively, that smaller, higher-density lots had less surface heat island formation than larger, lower-density lots, which tended to have more vegetative cover after controlling for the number of bedrooms per home, or the number of people the home was designed to accommodate. The amount of lawn and landscaping in an area, which is strongly correlated with lot size, was actually a stronger predictor of surface heat island formation in urban areas than the amount of impervious cover, although

Exhibit 4-9: Cooling off in Centennial Olympic Park, Atlanta. The heat island effect can make cities uncomfortably hot in the summer unless there are places to cool off.
Photo source: Patrick Fitzgerald via flickr.com

both were important. A 25 percent reduction in impervious cover of a single-family lot was associated with a 16 percent reduction in surface heat island formation. If combined with a 25 percent reduction in

---

[427] Marshall, Brauer, and Frank 2009
[428] Marshall, McKone, Deakin, and Nazaroff 2005
[429] Frank and Engelke 2005
[430] Stone and Norman 2006

lawn and landscaping area (i.e., smaller lot sizes), the reduction in surface heat island formation reached 28 percent.

Research has also shown a link between the built environment and extreme heat events, which the heat island effect exacerbates. Over five decades, the number of extreme heat events in 53 U.S. metropolitan regions increased by about 0.2 events per year overall, leading to about 10 additional events per city in 2005 compared to 1956. [431] A comparison across these metropolitan regions showed that the most sprawling (the top quartile based on a sprawl index) had a rate of growth in heat events more than twice that of those in the bottom quartile. The most compact metropolitan regions had an average of 5.6 more extreme heat days in 2005 compared to 1956, while the least compact had 14.8 more days, after controlling for climate zone, metropolitan population size, and the rate of metropolitan population growth. Lower rates of tree canopy loss in more compact areas might contribute to this difference. The study found that between 1992 and 2001, the most sprawling metropolitan regions had twice the rate of tree canopy loss as the most compact regions.

Water Quality

Compact development also influences water quality. Section 3.3 discussed the impacts of development and its associated impervious cover on downgradient[432] water resources and water quality. Since compact development means less land is needed for a given number of people, it can play an important role in protecting land from degradation and preserving downgradient water quality.

A model to evaluate pollutant loadings in stormwater runoff under different residential densities showed that as the number of homes per acre increases, so do the total amount of stormwater runoff per acre and the total pollutant loadings per acre, including total nitrogen, total phosphorous, and total suspended solids.[433] However, the model also showed that 100 homes developed at a density of eight or more per acre produced less total stormwater runoff and pollutant load than 100 homes developed at a typical suburban density of three to five homes per acre. In other words, for a constant number of households, denser development generates less total stormwater runoff because it affects a smaller area. EPA found similar results in a modeling study that showed that for a given amount of development, higher-density development produces less runoff and less impervious cover and affects less of the watershed than low-density development.[434]

4.2.2   Mixed-Use Development

Density alone might be less important than the mix of uses in an area, which affects distances between destinations and the ability to walk or bike between them. Standard zoning separates uses into distinct zones for residential, commercial, or industrial uses. In contrast, mixed-use development puts land uses with complementary functions close together. Complementary uses include housing, shopping and

---

[431] Stone, Hess, and Frumkin 2010
[432] Downgradient refers to the direction that groundwater flows.
[433] Jacob and Lopez 2009
[434] EPA, *Protecting Water Resources with Higher-Density Development* 2006

entertainment, offices, restaurants, schools, and houses of worship—any destinations to which people regularly travel.

When an office building also contains shops and restaurants, the infrastructure that supports the building—the roads and parking lots—is used for more of the day. Office workers use the parking lot during the day. Restaurant and theater patrons use the parking in the evenings. The alternative is two sets of roads and parking lots—one set serving office buildings and another serving retail and entertainment areas.

Exhibit 4-10: Main Street in Montpelier, Vermont. A mix of shops, offices, and apartments in the city's downtown create a neighborhood where residents are close to a lively arts and music scene, restaurants, and businesses.
Photo source: EPA

Mixed-use development can occur on different levels: site-specific, neighborhood, or regional. On a site, individual buildings or complexes can incorporate a variety of uses. For example, a single building might include apartments, offices, and shops. At the neighborhood level, mixed-use development refers to the arrangement of different uses across several blocks or acres of land so that they are connected. At the regional level, mixed-use policies often aim to balance jobs and housing so that people can live closer to their places of employment.

At any level—building, neighborhood, or region—the travel-related environmental effect of mixing uses is similar. By making it easier for people to walk, bike, and use transit to reach destinations, mixed-use development patterns allow people to drive less if they choose. Reductions in VMT can lead to decreases in automobile emissions, thereby improving air quality.

Mixing land uses can reduce VMT in several ways:

- **Trip lengths**—By putting destinations closer together, a mix of land uses can minimize travel distances and improve access to jobs, services, or recreation.

- **Mode choice**—Putting destinations closer together allows trips to be made by walking and bicycling rather than by driving cars. Even if people drive to one destination, they can walk, bike, or use transit to get to another nearby. Mixing jobs and housing in an area can reduce commute distances, making walking, biking, and transit more practical. Alternatively, mixing jobs or homes with shops and businesses lets workers take care of errands during the day, without needing a car.

- **Vehicle ownership**—Easy walking and biking access to jobs and shopping reduces the need to own a car to meet daily needs, reducing VMT.

As noted in Section 4.2.1, much of the literature on compact development shows that the effect of density in reducing VMT is much stronger when combined with other characteristics, including mixed land uses.[435] Fewer studies have attempted to isolate the effects of a mix of uses, as a factor influencing VMT, from other characteristics with which it is generally associated, such as having multiple destinations easily accessible from around an area.[436] In one example, a study of Portland, Oregon, neighborhoods showed that the more diverse the land use mix, the lower the rates of commuting by driving alone. The effect was stronger for mixed-use development in residential areas than job centers.[437]

Researchers analyzing data from the San Francisco Bay Area found that every 10 percent increase in the number of jobs that are in the same occupational category as a person currently works and that are located within 4 miles of that person's home was associated with a 3.3 percent reduction in commuting VMT. Every 10 percent increase in the number of retail and service jobs within 4 miles was associated with a 1.7 percent reduction in VMT for shopping and services.[438] Factoring in estimates of the amount of VMT attributable to commuting (37 percent) versus shopping and services (43 percent) in the San Francisco Bay Area, improved access to jobs in the same occupational category as the job one currently holds was associated with a 73 percent greater reduction in VMT than improved access to shopping and services.[439]

In a more comprehensive analysis, researchers studied 239 mixed-use developments across six U.S. metropolitan regions for which good household travel data were available: Atlanta; Boston; Houston; Portland, Oregon; Sacramento, California; and Seattle. The results showed that on average, three out of every 10 trips in a mixed-use development are relatively short and remain entirely within the development, adding no traffic to surrounding streets. Mixed-use developments that had the largest share of internal versus external trips had the greatest diversity of activities available within the development and/or were in walkable areas with good transit access. The more centrally located the development is and the more jobs are close by, the lower the overall VMT of residents.[440] Other research has shown that some portion of the trips occurring within mixed-use developments likely would not have occurred if not for their relative ease.[441] Thus, the trip reduction benefits of mixed-use development might be overestimated if not accounting for this effect. However, mixed-use development has benefits beyond reducing VMT. Any additional trips in mixed-use communities are likely caused by and in turn contribute to a vibrant pedestrian atmosphere.

---

[435] National Research Council of the National Academies, *Driving and the Built Environment* 2009
[436] Kuzmyak, et al. 2003
[437] Jun 2008
[438] VMT for shopping and services included any trip that involved travel to at least one shopping, service, or eating destination that did not also include a work destination. Both VMT and vehicle hours traveled included only time spent in a personal vehicle.
[439] Cervero and Duncan 2006
[440] Ewing, Greenwald, et al. 2011
[441] Sperry, Burris, and Dumbaugh 2012

A review of the literature found a consistent positive relationship between the amount that people walk[442] as a means of transportation and population density, the distance to non-residential destinations, and the degree to which land uses are mixed, all of which suggest the importance of destination proximity in explaining walking behavior.[443] A more recent study of an ethnically diverse sample of 5,529 adults from six U.S. cities found that when population density and the amount of land devoted to retail uses increases from the fifth percentile to the 95th percentile, the probability of walking more than 150 minutes per week, compared to getting no exercise, increased from 66 percent to 95 percent.[444] Other research suggests that the proximity of specific land uses is likely also important for determining the amount that people walk. A study in Montgomery County, Maryland, found that in addition to land use diversity, the number of bus stops, grocery stores, offices, and retail stores within half a mile from home were significant predictors of how likely people were to walk to get to their daily destinations.[445]

4.2.3 Street Connectivity

Many communities contain a hierarchy of dead-end or cul-de-sac local streets with strictly residential uses that lead to collector streets where all retail and commercial activity occurs (Exhibit 4-11). These collector streets lead to major arterials that connect communities to others via highways. Some communities are bounded by railroad tracks, lakes, or other physical barriers, and some do not have sidewalks. These patterns make pedestrian and bike travel difficult because circuitous routes and limited access increase trip length. Collector and arterial streets tend to be wide to allow vehicles to move faster and to

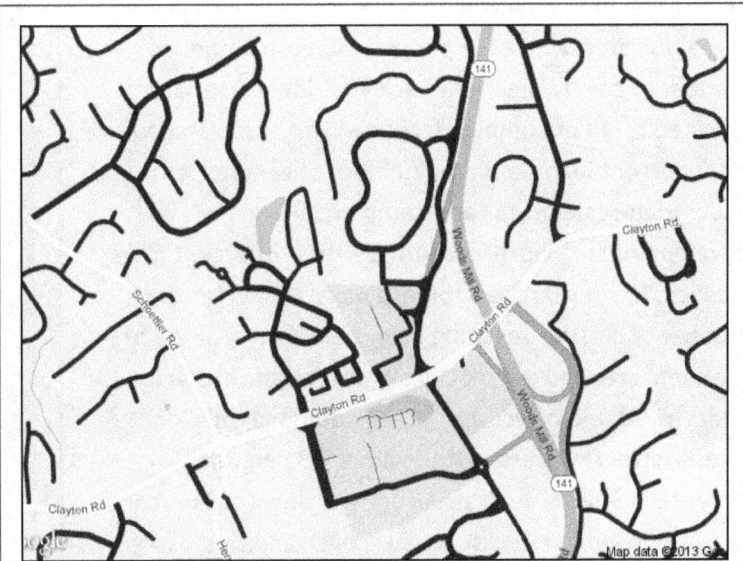

Exhibit 4-11: Map of Chesterfield, Missouri. Neighborhood roads (in dark blue) serve only local residents, directing all traffic to arterial streets (in yellow) where most retail is located. Highways (in red) also serve local travel needs, but carry large volumes of traffic at high speed.
Source: © 2013 Google

handle the large traffic volumes channeled to them from smaller neighborhood streets. Wide streets are difficult and often dangerous for pedestrians and bicyclists to cross or to share with vehicles, especially if

---

[442] Studies included in the review used a variety of different metrics to measure the amount people walked, some of which were based on time spent walking and others on the number of trips.
[443] Saelens and Handy 2008
[444] Rodriguez, et al. 2009
[445] McConville, et al. 2011

they lack sidewalks or crosswalks. Such poor pedestrian environments discourage walking and bicycling, leading people to rely on driving, even for short trips. Street grids with short blocks can provide multiple routes for traffic and make walking and biking easier and safer.

Many communities created hierarchical street patterns in the belief that widening and straightening streets, eliminating intersections, and reducing neighborhood traffic volumes by locating retail along arterials can improve traffic safety.[446] However, traffic safety studies have failed to support this belief. For example, an analysis of the association between car crashes and urban form in the city of San Antonio, Texas, found that locating retail and commercial uses on arterial streets away from residential areas and designing roads to funnel all traffic through an area on these arterials made streets more dangerous.[447] Each additional mile of arterial street was associated with an increase of 15 percent in the number of car crashes and a 20 percent increase in the number of fatal crashes; each additional large, single-use store[448] was associated with a 6.6 percent increase in the number of car crashes; and each additional arterial-oriented retail or commercial parcel was associated with a 1.3 percent increase in the number of car crashes. Pedestrian-scaled retail and commercial developments[449] were associated with 2.2 percent fewer crashes. The researchers found similar trends for the number of crashes resulting in injuries. A later study of the same area looked specifically at crashes involving pedestrians and bicyclists.[450] Each additional mile of arterial street was associated with a 9.3 percent increase in vehicle-pedestrian crashes and a 6.6 percent increase in vehicle-cyclist crashes; each additional large, single-use store was associated with a 8.7 percent increase in vehicle-pedestrian crashes; and each additional arterial-oriented retail or commercial parcel was associated with a 3 percent increase in vehicle-pedestrian crashes and a 1.7 percent increase in vehicle-cyclist crashes.

Exhibit 4-12: Downtown Fredericksburg, Virginia. Short blocks create a compact street grid in the city's downtown retail district, making walking easy and enjoyable. Photo source: EPA

---

[446] Dumbaugh and Rae 2009

[447] Dumbaugh and Rae 2009

[448] A large, single-use store was defined as a parcel in a retail use, with a single building occupying 50,000 square feet or more, and having a floor-area ratio of 0.4 or less. Stores of this design tend to have large surface parking lots and attract traffic from a broad geographic area.

[449] Pedestrian-scaled retail and commercial developments were defined as parcels whose retail or commercial uses occupy buildings of 20,000 square feet or less and had floor-area ratios of 1.0 or greater.

[450] Dumbaugh and Li 2010

Street connectivity affects not only traffic safety but also environmental impacts. A meta-analysis of travel literature showed that the degree of street connectivity is associated with the amount of car use compared to walking and transit use.[451] A study of 45 of the largest U.S. metropolitan regions considered street connectivity, as measured by a composite factor that incorporated average block length in the urbanized portion of the metropolitan area, average block size, and the percentage of small blocks.[452] Each standard deviation increase in connectivity was associated with a reduction of approximately 5.5 high-ozone days per year. These results are likely due to decreased vehicle emissions as well as decreased heat island effect (see Section 3.5). A study of census block groups in King County, Washington, looked at the correlation between a walkability index,[453] VMT, and air quality.[454] It found that a 5 percent increase in walkability was associated with a 6.5 percent reduction in VMT, a

5.6 percent reduction in nitrogen oxide emissions, and a 5.5 percent reduction in emissions of VOCs. Researchers analyzed data from a 12-county region in North Carolina and found that a 10 percent increase in the number of intersections in a neighborhood was associated with a 4 percent reduction in daily travel distance.[455]

4.2.4   Community Design

When communities are designed only to make car travel fast and easy, they can discourage walking and biking by making it dangerous, inconvenient, and unpleasant. Aspects of the built environment such as the number of off-street paths between destinations for cyclists and walkers, the design of streets and parking, building orientation, and building design all contribute to the relative appeal of that area to pedestrians and bicyclists and to the

Exhibit 4-13: Pedestrian-friendly streetscape. Marked crosswalks and curb extensions into the street help make Dickson Street in downtown Fayetteville, Arkansas, safe and enjoyable to walk on.
Photo source: EPA

---

[451] Ewing and Cervero 2010

[452] Stone 2008

[453] The walkability index was based on the number of intersections per square kilometer, the number of residential units divided by acres in residential use, land use mix, and retail floor-area ratio (retail building floor area divided by retail land area). The number of intersections was given twice the weight of the other three variables.

[454] Frank, et al. 2006

[455] Fan and Khattak 2008

area's general aesthetic appearance. Together, these are often referred to as "microscale" urban design factors—small-scale elements that affect the safety, convenience, and desirability of living and working in more compact areas and of using transit, walking, and bicycling.

Parking is one important component of community design. The amount, type, and placement of parking determine not only how convenient it is to drive to destinations, but less obviously, how convenient it is to walk or bike once there. Vast areas of parking create distances between destinations that are less practical and less appealing to traverse by foot or by bike.[456] In addition, having parking space for more cars than nearby streets are able to accommodate tends to exacerbate congestion.[457] The amount of space cities devote to parking varies considerably. An analysis of the number of parking spaces per job in the central business district of several cities in the United States found that values ranged from 0.06 in New York City to 0.91 in Phoenix.[458]

Street design is another component of community design with important effects on travel mode choice. In areas that do not include adequate bicycle and pedestrian facilities such as sidewalks, bike lanes, and crosswalks, travel by foot or bike can be intimidating and unpleasant. By making walking more desirable than driving, urban design factors can encourage more walking or bicycling trips, which can reduce vehicle travel and emissions if those trips replace vehicle trips.

Features that improve the pedestrian environment include sidewalks, clearly marked crosswalks and walk signals, lighting, and other amenities such as shade trees, benches, and streetscapes designed with the pedestrian in mind. Features that improve the bicycling environment include bicycle paths and lanes on streets, bicycle parking, and clear signage for all of these. Communities with streets designed for the safety of all users, also known as *complete streets,* can facilitate walking and biking and help residents lead healthier lifestyles.[459] For example, a review of the literature identified six studies on the effectiveness of street and sidewalk improvements designed to increase physical activity.[460] Overall, the median increase in the number of people walking or biking due to these improvements was 35 percent. An analysis of the applicability of the results found diverse geographic locations and populations are likely to respond similarly. One study on 90 of the largest 100 U.S. cities found that those with the most bike lanes per resident (those in the highest quartile) had rates of bike commuting three to four times those of cities with the fewest bike lanes (those in the lowest quartile).[461] A review of literature on the effectiveness of a wide variety of strategies used worldwide to increase bicycle use concluded that individual interventions, such as adding bike lanes or other bike infrastructure, are likely to be more effective when implemented as part of a comprehensive strategy—including, for example, programs and land use planning to encourage bike use.[462]

---

[456] Mukhija and Shoup 2006
[457] Manville and Shoup 2005
[458] Manville and Shoup 2005
[459] Giles, et al. 2011
[460] Heath, et al. 2006
[461] Buehler and Pucher 2012
[462] Pucher, Dill, and Handy 2010

Many communities eliminated—or never provided—pedestrian and bike amenities because they believed that streets that are relatively forgiving of driver error are the safest. However, a review of the literature concludes that in areas with a lot of development, street design elements that slow down drivers, such as narrow lanes, traffic-calming measures, and street trees, actually make roads the safest.[463] For example, an evaluation of 13 road safety measures implemented in New York City between 1990 and 2008 found that reducing the number of travel lanes, along with adding left-turn lanes (and sometimes bike lanes) reduced the number of crashes at intersections by 13 percent and along road segments by 67 percent.[464] Other measures that

Exhibit 4-14: 9th Avenue bicycle lane in New York City. A dedicated bike lane with separate traffic signals separates bicyclists from vehicles on city streets.
Photo source: Kyle Gradinger/BCGP via flickr.com

reduced crashes included installing new signals (25 percent) and adding a left-turn phase to existing signals (17 percent). Pedestrian crashes were also reduced by an all-pedestrian signal phase that stops traffic in all directions (35 percent), increased pedestrian crossing time (51 percent), and high-visibility crosswalks (48 percent). This study found that curbside bus lanes actually increased the total number of crashes,[465] and adding bike lanes increased the number of bicycle crashes at intersections. However, the bike lanes evaluated did not include special accommodations for bicyclists at intersections, and the study did not account for the possibility that more people were bicycling after the bike lanes were added. A review of studies that investigated the effect of bicycle-specific infrastructure on bicyclist safety found that clearly marked bike routes and lanes cut injury or crash rates roughly in half compared to on-road bicycling with traffic and also consistently improved injury or crash rates compared with off-road bicycling on routes shared with walkers and other users.[466]

---

[463] Ewing and Dumbaugh 2009
[464] Chen, et al. 2013
[465] The city installed bus lanes to improve bus speed and attract riders, not as a safety measure. The authors hypothesize that the unexpected safety decline from bus lanes could be due to a lack of enforcement; cars frequently blocked the lanes and required buses to go around them. In addition, the bus lanes evaluated were not painted as are newer bike lanes in the city to improve visibility.
[466] Reynolds, et al. 2009

Beyond affecting travel mode choice, street and parking lot design can also have direct water quality impacts. Designers can integrate elements to store, infiltrate, and evapotranspirate water into streets, alleys, and parking areas, helping to minimize stormwater runoff (see Section 3.3). The design elements of green streets and green parking include permeable pavements, vegetated swales, planters, and trees.[467] (See Section 4.2.7 for discussion of other site-scale techniques to manage stormwater runoff.) An evaluation of the performance of four commercially available permeable pavement systems after six years of use found that they infiltrated essentially all precipitation during the three-month study period in Renton, Washington, outside of Seattle.[468] The quality of the water that moved through the permeable pavement was better than that of the runoff leaving an asphalt parking area. Both copper and zinc concentrations were below toxic levels and usually below detectable levels in all infiltrated stormwater, while asphalt runoff consistently contained copper and zinc at toxic concentrations. Another study of permeable pavements in North Carolina found that they can reduce and sometimes even eliminate stormwater runoff.[469] Of two sites monitored for water quality, only one had any discharge during the study period. Its discharge of total nitrogen, total phosphorus, ammonia, and zinc was lower than from adjacent asphalt lots.

4.2.5   Destination Accessibility

The form of our built environment also determines how accessible different places are to each other on a regional level. While compact development, a mix of land uses, and good community design are important factors at a local level, researchers have attempted to define indicators of overall *destination accessibility* (how easy it is to access attractions across the region) that are not directly captured by these other metrics.

A meta-analysis of travel literature available at the end of 2009 found that destination accessibility is the measure of the built environment that is most strongly associated with VMT and the amount people walk.[470] The authors suggest that these results show any centrally located development is likely to generate less car traffic than a remotely located development even if the remote development is compact, mixed-use, and designed for pedestrian travel.

Researchers have also studied how *neighborhood accessibility* influences travel behavior. Neighborhood accessibility refers to the ease of traveling within and between neighborhoods by a variety of options, including car, transit, walking, and biking. It is based on measures of population density, land use mix, and average block size. A study measured how the travel behavior of 430 households changed when they moved between locations with different levels of neighborhood accessibility.[471] It found that households relocating to areas with higher neighborhood accessibility reduced their VMT, person miles

[467] EPA, *What is Green Infrastructure?* n.d.
[468] Brattebo and Booth 2003
[469] Bean, Hunt, and Bidelspach 2007
[470] Ewing and Cervero 2010
[471] Krizek 2003

traveled, and the number of destinations per trip while simultaneously increasing the number of trips taken.

Another study considered the factor of *urban contiguity* and its relationship to air pollution levels.[472] Contiguity measures the patchiness in a developed area, or how much development occurs isolated from other developed areas.[473] Satellite measurements of nitrogen dioxide levels in 83 cities found that urban contiguity is associated with lower levels of nitrogen dioxide. Specifically, a decrease in contiguity of one standard deviation is associated with a 31 percent increase in nitrogen dioxide levels. While population size was the single most important predictor of nitrogen dioxide levels in a city, results showed that nitrogen dioxide increases caused by a 10 percent increase in population could be offset by a 4 percent increase in urban contiguity.

Finally, one study showed that air quality is also related to *population centrality*, a measure of how a population is distributed relative to a central business district. The study found that while population density was associated with higher population-weighted levels of $PM_{2.5}$ and aggregate pollutant levels, population centrality was associated with lower population-weighted levels of these measures as well as ozone.[474] The authors speculate that these findings could be explained in part by how travel patterns are influenced by population centrality and by how air pollution varies across a geographic area. Ozone concentrations tend to peak at some distance from an urban center, so as populations are more dispersed, more people are located in this zone of highest concentration.[475]

4.2.6   Transit Availability

In addition to locating development near transit stations, how we build and manage transit and related infrastructure determines how useful and convenient public transit is, which in turn influences how much people choose to use it. Transit access can be improved not only by how we build communities surrounding transit, but also by expanding the supply of transit itself through construction or service improvements (Exhibit 4-15). For example, among a set of residents surveyed before and after a new light-rail stop opened in a Salt Lake City neighborhood, ridership increased from 50 to 69 percent.[476] Service improvements that can improve ridership include increasing frequency, particularly on routes with infrequent service; making service more reliable; having easy-to-remember departure times and readily available schedules; expanding routes, particularly to unserved or poorly served areas; adding new buses; and lowering fares.[477,478] Researchers developed a model to study the factors affecting the

---

[472] Bechle, Millet, and Marshall 2011
[473] Contiguity is calculated as the ratio of the main contiguous built-up area to the total built-up area of a city. It is not correlated with compactness, calculated as the ratio of built-up area to total buildable area within a circle surrounding the main built-up area of a city.
[474] Population centrality is not correlated with population density.
[475] Clark, Millet, and Marshall 2011
[476] Brown and Werner 2007
[477] Evans 2004
[478] Pratt and Evans 2004

widely varying levels of transit ridership across U.S. cities.[479] They found that four types of factors can explain most of the variation:

Exhibit 4-15: Streetcar in San Francisco. The F Market & Wharves line operates 20 hours a day, 365 days a year, connecting destinations popular with both commuters and tourists. Service runs at least every 15 minutes and as frequently as every 6 minutes.
Photo source: EPA

- **Regional geography** (i.e., total population, population density, geographic land area, and regional location).
- **Median household income**.
- **Population characteristics** (i.e., political party affiliation and percentage of households without a car).
- **The number of trips taken using neither transit nor a personal vehicle** (e.g., walking, biking, or carpooling).

While these types of regional characteristics are most influential in determining levels of transit use, transit policies are important as well. This study found that more frequent service and lower fares can at least double transit use in a given area.

The effects of increased transit service on VMT are less clear. A study of 228 U.S. metropolitan areas found that the number of buses and train cars in a region does not influence VMT.[480] However, models have shown that modest effects are possible; transit service improvements can reduce VMT by up to 1 percent during the first 10 years after investment and up to 2 percent thereafter.[481] In Portland, Oregon, neighborhoods with more public transit (more bus stops and/or shorter distances from homes to light rail stations) have fewer people who commute by driving alone.[482] Data on air pollution support these findings. One study found that the availability of transit is associated with lower population-weighted concentrations of $PM_{2.5}$.[483]

Irrespective of its effect on VMT, transit availability can also improve users' health. Transit users who walk to their transit stop spend on average 24 minutes per day walking for travel, and 29 percent of these transit users meet the U.S. surgeon general's recommendation of at least 30 minutes of physical

---

[479] Taylor, et al. 2009

[480] Duranton and Turner 2011

[481] Rodier 2009

[482] Jun 2008

[483] Clark, Millet, and Marshall 2011

activity per day by walking to and from transit.[484] A study of Charlotte, North Carolina, residents before and after construction of the city's light rail system found that daily use of the system was associated with a reduction in body mass index[485] and an 81 percent reduction in the likelihood of becoming obese over a 12- to 18-month period.[486]

4.2.7   Green Building

How we build our communities involves not just the layout of streets and buildings, but also the materials and approaches we use to construct the buildings. The term *green building* refers to the practice of using resources efficiently and reducing impacts on human health and the environment in all phases of a building's lifecycle: pre-design, siting, design, construction, operations, maintenance, renovation, and demolition. Many of the decisions that developers make during the pre-design and siting phases have already been covered in other sections. For example, the location of a home helps determine how far people travel to commute to work, get to school, shop, and participate in recreational activities, as well as whether they travel by car, public transit, walking, or biking. The travel behavior of a building's occupants is an important component of a building's overall energy use, and part of making a building energy efficient is siting the building in a central location near businesses, schools, recreational options, and transportation options.

Although part of building green involves carefully choosing the site location, this section discusses other aspects of green building:

- Ensuring that building construction and renovation practices limit environmental impacts.
- Conserving energy and water in building operations.
- Using building materials that are environmentally safe for occupants.
- Designing sites to allow the capture and reuse or infiltration of stormwater.

These broad categories include just some of the wide range of techniques and strategies encompassed by green building approaches. This section discusses these approaches to provide a general sense of how green building could protect the environment and human health. Unlike approaches to how we build our communities, the most environmentally preferable products for constructing and operating buildings might vary depending on location and are likely to change as more performance data are collected and new products come on the market. In addition, rather than simply adopting individual practices from a menu, the most successful green building projects use an integrated, interdisciplinary approach that considers how individual components will interact with each other and with the chosen site. Teams including architects, construction contractors, operations staff, and building occupants collaborate to plan integrated systems and optimize performance after building construction.

---

[484] Besser and Dannenberg 2005. Median walking time was 19 minutes.
[485] Body mass index for light rail users compared to comparable non-users declined by 1.18 kilograms per meter squared, or 6.45 pounds for a person who is 5 feet, 5 inches.
[486] MacDonald, et al. 2010

Exhibit 4-16: Tupelo Alley apartments in Portland, Oregon. This building was awarded a gold certification by Leadership in Energy and Environmental Design (LEED) for its water and energy efficiency and the use of low-emitting materials inside to provide better indoor air quality. It is also within walking distance of neighborhood amenities and an easy commute by bike or transit to downtown.
Photo source: EPA

The National Academy of Sciences estimated in 2009 that 57 million new housing units would be needed by 2030 to accommodate projected population growth.[487] This estimate does not include additional buildings needed for commercial and industrial uses. The choices we make about how to construct these buildings will be important determinants of the ecological and human health impacts of our built environment. In addition, many green building techniques are applicable to building retrofits and renovations, so the potential for better environmental performance from the building sector is large indeed.

Construction/Renovation Practices

Green building techniques involve a wide range of construction practices designed to minimize the environmental and health impacts of construction. Advances in construction technology and research evaluating impacts of different options will change our understanding of best practices. This section presents a sample of green building construction practices to illustrate the issues they can address.

One green building strategy is using locally or regionally sourced construction materials to shorten the distance between the suppliers and construction sites, thereby reducing transportation emissions. Other strategies include limiting sediment and nutrient releases from construction sites, preventing soil erosion, limiting the amount of land disturbed, designing buildings to facilitate their deconstruction at the end of their useful life, and using energy and water more efficiently during construction.[488] For example, diesel engines in construction vehicles such as backhoes, bulldozers, excavators, and loaders emit air pollution, including NO_X, particulate matter, hydrocarbons, and carbon monoxide.[489] Several strategies can reduce construction vehicle emissions, including reducing idling time, improving maintenance, purchasing energy-efficient models, and using technological controls to reduce emissions.[490]

---

[487] National Research Council of the National Academies, *Driving and the Built Environment* 2009
[488] Shen, et al. 2007
[489] Lewis, et al. 2009
[490] Lewis, et al. 2009

Radon-resistant construction techniques for mitigating radon exposure can involve both passive methods, such as improving ventilation to the outside or sealing radon entry points, and active methods such as sub-slab depressurization, which draws radon out of the sub-soil to the outside air. The best method will depend on building type, soil conditions, and climate, but radon reductions of up to 90 percent are possible.[491]

Energy Efficiency

Green building strategies can improve the environmental performance of buildings by making them more energy efficient. Homes built between 2000 and 2005 were 14 percent more energy efficient per square foot than homes built in the 1980s and 40 percent more energy efficient than homes built before 1950.[492] More stringent residential building energy efficiency codes and standards are partly responsible for this improvement.[493] However, through various green building techniques and strategies, a building's energy efficiency can well exceed that required by building codes, and the potential to improve the energy consumption of the building sector (see Section 2.2.2) through both new construction and renovation is significant. A review of retrofits of commercial buildings found energy savings of 50 to 70 percent around the world.[494] For example, converting the heating systems in 10 schools from low-pressure steam systems to low-temperature hot water systems reduced heating energy use by an average of two-thirds.[495]

Exhibit 4-17: Cherokee Mixed-Use Lofts in Hollywood, California. This residential and retail building is 40 percent more efficient than required by California's building code; has a green roof, water-efficient fixtures, drought-tolerant landscaping, and solar heating; and uses materials throughout that are recycled and contain low or no VOCs. It is close to shops, services, and other amenities so residents can walk for many daily needs.
Photo source: Calderoliver via Wikipedia Commons

Many different techniques and strategies can help improve the energy efficiency of buildings:

- **Insulation**—Effective insulation and good windows create a barrier between indoor and outdoor air, limiting heat loss or gain, depending on the season. In areas with cold climates, houses built to maximize the thermal barrier between indoor and outdoor air use only 10 to

---

[491] Rahman and Tracy 2009

[492] U.S. Department of Energy 2012

[493] The 2006 International Energy Conservation Code requires buildings to be 14 percent more energy efficient than the original code in 1975.

[494] Harvey 2009

[495] Durkin 2006

25 percent of the energy of a house built to minimum code standards.[496] Energy-saving strategies can also reduce the amount needed for cooling by more than 50 percent. In areas with cool evenings, allowing adequate nighttime ventilation can often eliminate the need for air conditioning entirely.[497] As the climate warms and extreme heat events increase, better-insulated homes can help people avoid heat-related illnesses and death by reducing the amount of energy they need to use to keep their home at a safe temperature.

While improving the thermal performance of buildings, green building practices also aim to control the amount of indoor air moisture to prevent the growth of mold and avoid allergic and respiratory effects (discussed in Section 3.4.4). Controlling both insulation and air leakage is critical for achieving an appropriate moisture balance, particularly for retrofits of older homes where it can be more difficult.[498]

- **Heating and cooling systems, appliances, and lighting**—An optimal heating, ventilation, and air conditioning (HVAC) system can reduce energy use by 30 to 75 percent, in addition to any savings achieved by other techniques.[499] EPA and the U.S. Department of Energy created the voluntary ENERGY STAR program in 1992 to help consumers and businesses identify the most energy-efficient products, homes, buildings, and practices.[500] An evaluation of the impacts of the program found that from 1992 through 2006, the use of ENERGY STAR-labeled products, including office equipment, consumer electronics, residential HVAC systems, lighting, and appliances, saved 4.8 exajoules (10^{18} joules) of energy and avoided 82 teragrams of carbon dioxide equivalent.[501] This savings represents about 1.5 percent of the 311 exajoules consumed by the entire residential sector between 1992 and 2006 (see Exhibit 2-11). The savings came primarily from computer equipment, residential light fixtures, televisions, and furnaces. The same study projected that between 2007 and 2015, ENERGY STAR-labeled products would save 12.8 exajoules of energy and avoid 203 teragrams of carbon dioxide equivalent.[502]

Energy-efficient lighting involves not only using better technology, but also minimizing the need for artificial lighting through *daylighting* techniques. Incorporating skylights, solar tubes, and north-facing windows into the building design, as well as positioning spaces and furniture to maximize sunlight that enters the structure, can illuminate the interior space without consuming electricity.[503] Minimizing artificial lighting also reduces excess heat from light bulbs, which in turn lowers the energy demand for air conditioning. For example, in perimeter offices,

---

[496] Harvey 2009
[497] Harvey 2009
[498] Lubeck and Conlin 2010
[499] Harvey 2009
[500] EPA, *About Energy Star* n.d.
[501] Sanchez, et al. 2008
[502] Sanchez, et al. 2008
[503] U.S. Department of Energy 2002

daylighting can reduce energy used for lighting by 40 to 80 percent and energy used for lighting and cooling together by 20 to 33 percent.[504]

- **Passive strategies**—Passive strategies involve modifying key design elements, including building size, orientation, the height-to-floor area ratio, and the wall-to-window area ratio. In general, the larger the building is, the more energy needed for heating and cooling. For example, researchers estimate that if 1 percent of U.S. households lived in a 2,000-square-foot single-family house rather than one that is 2,400 square feet, the United States would save 3,164 billion Btu annually.[505] Savings are more significant

Exhibit 4-18: Sustainably designed kitchen. This kitchen in Eagle Rock, California, uses natural lighting to reduce the need for electricity. Rolling wood screens can be moved to shade the south-facing windows during hot weather, and a tree grows through the outdoor deck to provide additional shade. Sustainable materials were used in the construction, including countertops made with recycled coal fly ash.
Photo source: Jeremy Levine Design via flickr.com

when comparing single-family homes to homes in multi-unit buildings because of the efficiencies gained by shared walls, ceilings, and floors. If 1 percent of households lived in a 2,400-square-foot apartment rather than a single-family home of the same size, researchers estimate 27,906 billion Btu could be saved annually. Other passive strategies include using natural ventilation to reduce the energy needed for cooling and orienting buildings on a site to take advantage of the optimal position relative to the sun, wind direction, and topography.

These approaches are just some of the energy-saving strategies and technologies for green building design and operations. A review of the literature shows that using combinations of available options could reduce annual energy use per unit of floor area by a factor of three to four for new buildings and two to three for existing buildings.[506]

Water Conservation

Green building strategies can increase the environmental performance of buildings by using water more efficiently. Not only does reducing water consumption conserve a valuable resource, it reduces the energy needed to pump water from a water treatment plant to homes, to heat the water (in some

---

[504] Harvey 2009
[505] Kockelman, et al. 2009
[506] Harvey 2009

cases), to pump it back to a wastewater treatment plant, and finally, to treat it before discharging it back to the environment.

Water-efficient household appliances and fixtures can yield significant water savings. For appliances using hot water, energy savings can also be substantial. An assessment of standard versus efficient clothes washers, toilets, and showerheads, found that efficient models used 38 to 58 percent less water, consume 28 to 35 percent less energy, and emit 28 to 52 percent fewer greenhouse gases over their entire lifecycle, including manufacture, use, and end-of-life disposal.[507] In a similar study, researchers analyzed the potential reductions of water use, energy use, and greenhouse gas emissions from using available water-efficient clothes washers, dishwashers, faucets, and showerheads. They found that in Australian households, up to 10.2 kiloliters (2,695 gallons) of water and 965 kilowatt hours of energy per person per year could be saved, amounting to a 30 percent reduction in water use and a 60 percent reduction in energy use from these household appliances and fixtures.[508]

Outdoor water use accounts for 31 percent of total water use for single-family households on average, but can be higher in arid climates (see Exhibit 2-16). Smaller lots with less turf or landscaped area need less water, and water-efficient landscaping and irrigation technology can significantly reduce water use. For example, drip irrigation delivers water only as fast as the soil is able to absorb it and limits surface water runoff and evaporation that often occurs with conventional sprinkler systems. Landscaping with native plants adapted to the natural rainfall patterns instead of using turf grass can also reduce water needs. In the arid western United States, estimates for how much changes in landscaping and irrigation practices could save range from 35 to 70 percent of per capita water use.[509] A study in Florida, where irrigation accounts for 64 percent of residential water use, found that changes in irrigation timing could reduce water use by 30 percent. When combined with replacing approximately half of the turf grass area with native plants, savings could reach 50 percent.[510]

Office and other types of commercial and institutional buildings such as hospitals, hotels, and schools can also achieve significant water savings using these and other techniques. Strategies for commercial and institutional facilities include developing a water management plan for each facility, regularly checking for and repairing leaks, using the most water-efficient bathroom fixtures, and optimizing cooling systems, including determining if they can provide or use on-site sources of water.[511]

EPA recognizes through its WaterSense labeling program[512] products that are at least 20 percent more efficient and perform as well or better than comparable products. The program also provides national specifications for water-efficient new homes and recognizes professional certification programs for landscape irrigation professionals that have verified proficiency in water-efficient irrigation system design, installation and maintenance, and auditing.

---

[507] Lee and Tansel 2012

[508] Beal, Bertone, and Stewart 2012

[509] Hurd 2006

[510] Haley, Dukes, and Miller 2007

[511] EPA, *WaterSense at Work* 2012

[512] EPA, *WaterSense* n.d.

Materials Selection

Careful selection of construction materials helps conserve natural resources and protect the environment. The building sector accounts for 24 percent of global natural resource extractions.[513] Using recycled, refurbished, and/or salvaged materials such as metal, glass, timber, brick, cement, and steel reduces the demand for raw materials; minimizes the amount of waste that needs to be disposed of; and reduces energy, chemical, and water use in manufacturing. Sturdy materials like concrete and steel, for instance, are energy-intensive to produce, so much so that the structural frame of a typical office building can account for 15 percent of its lifetime energy use.[514] A study of construction and demolition waste generation in Florida estimated that as much as 91 percent of building-related construction and demolition waste can be recycled with current technology, far more than the estimated 9 percent of construction and demolition waste that is actually recycled in the state.[515]

A lifecycle assessment that considered the energy footprints, water footprints, and contributions to global warming of different building materials found that they vary considerably.[516] For example, the study found that for exterior paving, clay tiles require 85 percent less energy and produce 66 percent fewer greenhouse gas emissions than ceramic tiles. For insulation, rock wool has an energy footprint four times lower and a water footprint 8.4 times lower and emits 4.7 times fewer greenhouse gases than polystyrene tiles and rigid polyurethane foam.

The materials selected for building construction affect indoor air quality. As discussed in Section 3.4, many of the materials that we use to build our homes and offices emit chemicals with known or potential adverse health impacts. Green building strategies seek to minimize or eliminate these materials. For example, researchers measured the total amount of several VOCs for several conventional and green materials.[517] All of the green materials emitted far fewer VOCs than their conventional counterparts, as shown in Exhibit 4-19.

Green Material	Total VOCs Emitted in 5 days (mg/m^2)	Conventional Material	Total VOCs Emitted in 5 days (mg/m^2)
Trex[518] decking wood	1.17	Pressure-treated wood	4.22
Ceramic floor tile	0	Vinyl floor tile	2.31
Water-based paint	18.9	Oil-based paint	6,940
		Wood stain	24,600

Exhibit 4-19: Total VOCs (in milligrams per square meter) emitted over a five-day period from green materials compared to conventional materials.
Source: James and Yang 2004

The materials selected for use inside buildings are also important for indoor air quality. EPA's Design for the Environment Program helps households, businesses, and institutions select cleaners and other

---

[513] Bribian, Capilla, and Uson 2011
[514] Horvath 2004
[515] Cochran, et al. 2007
[516] Bribian, Capilla, and Uson 2011
[517] James and Yang 2004
[518] Trex© is a wood alternative made of reclaimed and recycled wood and plastic fibers. This document does not convey official EPA approval, endorsement, or recommendation of this product.

products that are safer for the environment.[519] The program is based on a methodology for comparing the environmental and human health effects of chemical alternatives to minimize risk.[520]

A review of strategies to mitigate indoor air pollution exposures found that portable air cleaners can help reduce exposure to small airborne particles.[521] However, they are not proven effective for removing larger airborne particles (including many allergens) or VOCs, highlighting the importance of removing pollutant sources and ventilating indoor spaces with clean, outdoor air. Nevertheless, few scientific studies have evaluated the health benefits of avoiding products containing VOCs.

Site-Scale Green Infrastructure

Green infrastructure at the site scale, also known as low-impact development, is a strategy for managing stormwater where it falls, allowing soils and vegetation to absorb and filter the water, which reduces many of development's impacts on water quality that are discussed in Section 3.2. Examples of green infrastructure techniques include:

- **Infiltration** techniques, such as permeable pavements, disconnected downspouts, and rain gardens (Exhibit 4-20)—They are engineered structures or landscape features designed to capture and infiltrate stormwater, reduce runoff volume, and treat or clean runoff.

- **Evapotranspiration** practices, such as green roofs, bioswales, trees, and other vegetation—They can reduce stormwater runoff volumes by returning water to the atmosphere through evaporation of surface water or through transpiration from plant leaves. Trees and shrubs can also filter air pollutants and improve air quality.[522]

- **Capture and reuse** practices, such as rain barrels and cisterns—They capture stormwater for non-potable household uses, irrigation, or gradual infiltration.

Estimates of the performance of any given practice vary considerably because of wide variation in how and where green infrastructure is installed. For example, estimates for the runoff reduction from green roofs are between 50 and 100 percent, depending on the roof characteristics and annual precipitation patterns.[523] Researchers have consistently found potential building energy savings[524] and air pollution and carbon dioxide reductions[525] from green roofs, but precise values vary. Other environmental benefits of green roofs include the ability to neutralize acid rain, attract wildlife, and mitigate the heat island effect.[526] A review of the literature on the performance of individual green infrastructure practices found that bioretention areas, permeable pavements, and green roofs can reduce runoff volumes and

---

[519] EPA, *Design for the Environment* n.d.
[520] Lavoie, et al. 2010
[521] Sandel, et al. 2010
[522] Nowak and Greenfield 2012
[523] Rowe 2011
[524] Sailor, Elley, and Gibson 2012
[525] Rowe 2011
[526] U.S. General Services Administration 2011

Exhibit 4-20: Citygarden in St. Louis. Covering two city blocks in downtown St. Louis, Citygarden includes six rain gardens covering 5,000 square feet that capture stormwater runoff from the park and adjacent streets. Photo source: EPA

the concentration of many pollutants, including copper, lead, and zinc.[527] In 2009, the National Research Council reviewed recent studies on the performance of various bioretention techniques.[528] Runoff volume reductions ranged from 20 to 99 percent, with a median reduction of about 75 percent. The report concluded that practices that harvest, infiltrate, and evapotranspirate stormwater are important tools for reducing pollutant loadings from smaller storms. These smaller storms tend to carry away the bulk of the pollution on roads and parking lots. However, as a first line of defense, non-structural practices such as better designing building sites, disconnecting gutter downspouts, and conserving natural areas dramatically reduce runoff volumes and pollutant loadings resulting from development.

The benefits of green infrastructure are not just ecological. Green infrastructure can also make an area more attractive for residents and visitors and increase recreation space. In addition, a review of literature on the effect of green infrastructure on human health found that epidemiological, experimental, and survey data suggest that there is considerable potential for green infrastructure to improve the health and well-being of urban residents, likely due to physiological, emotional, and cognitive changes.[529]

4.3   Scenario Planning

Many of the patterns and practices discussed in this report—locating new development away from sensitive areas and on previously developed sites near transit; building compact, mixed-use communities; designing communities to serve bicyclists and pedestrians as well as cars; and using green building approaches—have demonstrated environmental benefits. Through these strategies, we can reduce land and habitat consumption, reduce the energy and resources needed for new infrastructure, and reduce growth in vehicle travel along with its associated human health and environmental impacts.

---

[527] Dietz 2007
[528] National Research Council of the National Academies, *Urban Stormwater Management in the U.S.* 2009
[529] Tzoulas, et al. 2007

As shown in Section 4.2, the efficacy of any one of these practices depends in large part on its context. The benefits of increasing housing or employment density, for example, depend on whether the density occurs along with a variety of easily accessible shops and services for residents and employees, whether streets are designed to make walking and biking between destinations convenient and safe, whether the development occurs near a central business district, and whether the buildings are designed to use energy and water efficiently. Using these strategies together enhances the benefits they bring individually.

This section considers several examples of scenario planning studies that look at how the combined effects of such land use strategies could improve the environmental outcomes of development. Scenario planning is the process of considering a range of plausible trends and evaluating the future outcomes that would likely result from each.[530] Scenarios are not forecasts or predictions of what will happen. They are intended to provide information for communities to understand how different land use decisions could lead to different outcomes based on assumptions about future development trends. Communities typically use scenarios to highlight key considerations for long-range planning or to understand the long-term impacts of short-term decisions. When a collaborative public process informs scenario planning and the modeling is based on locally important values, the results can help community leaders make broadly supported planning and development decisions.

Many regions have conducted scenario-planning exercises to help understand the anticipated impacts of land use decisions on the environment and human health. A 2007 review covered 80 scenario-planning exercises from more than 50 metropolitan regions.[531] For example, researchers developed four land use scenarios for a study of an area of southern California that includes north and west San Diego County and parts of Riverside and Orange counties. Each scenario had two variants: a population increase of 500,000 and a population increase of 1,000,000.

- The **Coastal Future** scenario concentrated residential development in high-density areas near the coast, away from rural areas.
- The **Northern Future** scenario concentrated new, low-density housing in suburban and rural areas of the northern part of the study area.
- The **Regional Low-Density Future** scenario spread large-lot development throughout the study area.
- The **Three-Centers Future** concentrated residential development near existing cities.

Modeling showed that the Three-Centers Future resulted in the best air quality outcomes for the region, while the Regional Low-Density Future resulted in the largest increase in VMT and the highest levels of air pollution.[532]

---

[530] Mahmoud, et al. 2009
[531] Bartholomew 2007
[532] Kahyaoglu-Koracin, et al. 2009

Long-term growth patterns in the eight counties of the San Joaquin Valley in California were also the subject of a modeling study that included:

- A **baseline growth** scenario that assumed no change in development trends for residential density.
- A **controlled growth** scenario that assumed road capacity would not increase, alternative forms of transportation would be expanded, and new residential growth would be high density.
- An **uncontrolled growth** scenario that assumed increases in road capacity, no expansion of alternative forms of transportation, and low- and very low-density new residential growth.
- An **as planned** scenario that assumed current plans—increases in road capacity, new high-speed rail, and no change in development trends for residential density—would be implemented.

The model projected that, regionwide, the controlled growth scenario would result in a 6 to 10 percent reduction in total emissions compared to the baseline growth scenario, while the unplanned growth scenario would result in a 7 to 10 percent increase in total emissions over the baseline growth scenario. Projections for reductions in VMT and vehicle emissions were the highest for areas that were relatively the densest at baseline.[533]

Another study assessed projected 2050 pollutant emissions across 11 major metropolitan regions in the Midwestern United States.[534] The study evaluated a business-as-usual scenario and a compact growth scenario, the latter of which was based on changes in the relative proportion of the population in urban, suburban, and rural census tracts between 1980 and 2000 in Portland, Oregon, a region that implemented several growth management policies over this time. The model projected VMT across all metropolitan areas to be 6.0 percent lower by 2050 under the compact growth scenario than under the business-as-usual scenario. Projections for vehicle pollutants took into account different travel speeds and frequencies of vehicle starts in urban, suburban, and rural areas. The model projected reductions of 6.0 percent for $PM_{2.5}$, 5.6 percent for NO_x, 5.6 percent for carbon monoxide, 5.3 percent for VOCs, and 5.1 percent for carbon dioxide under the compact growth scenario compared to the business-as-usual scenario.

A similar study of the same 11 major Midwestern metropolitan regions evaluated how development decisions could influence air quality changes projected to occur between 2000 and 2050.[535] Increased density was associated with a lower rate of VMT growth, although effects were more pronounced when density was increased in urban versus suburban areas. Overall, the model predicted VMT would increase 64 percent under the business-as-usual scenario and 56 percent and 47 percent under two different smart growth scenarios. The rate of growth in carbon dioxide emissions showed the same relative trends: 23 percent under the business-as-usual scenario and 17 percent and 15 percent under two different smart growth scenarios. Growth in carbon dioxide emissions was much lower than the VMT growth under all scenarios due to anticipated improvements in vehicle emissions technology. Put

---

[533] Niemeier, Bai, and Handy 2011
[534] Stone, Mednick, et al. 2007
[535] Stone, Mednick, et al. 2009

another way, under the smart growth scenarios, each 10 percent increase in population density was associated with a 3.4 percent reduction in the household VMT growth rate and a 3.0 percent reduction in the carbon dioxide emission growth rate.

The National Research Council published results of a national modeling exercise for different development scenarios:

- One scenario assumed that, beginning in 2000, 75 percent of all new homes would be built in more compact developments, with residents driving 25 percent less.
- A more moderate scenario assumed that 25 percent of new homes would be built in more compact developments, with residents driving 12 percent less.
- A base-case scenario assumed that all new homes would be built at the average density of homes built during the 1990s.

The committee estimated that by 2030, VMT, associated fuel use, and carbon dioxide emissions would be reduced below the base case by 7 to 8 percent for the first scenario and by about 1 percent for the second scenario. By 2050, reductions would be 8 to 11 percent and 1.3 to 1.7 percent, respectively.[536] The report noted that both scenarios are based on numerous assumptions that represent departures from current trends, and the authoring committee members disagreed about how realistic these assumptions are for the United States.

In another national modeling exercise, researchers evaluated five scenarios based on different levels of population density from 2005 to 2054: three in which population density declined by 47 percent, 39 percent, and 13 percent; one in which population density remained constant; and one in which population density increased 11 percent.[537] Under the scenario of increased population density, vehicle-related carbon dioxide emissions were projected to decline by 5 percent over 50 years compared to the scenario of no change in population density. Emissions were projected to increase by 15, 70, and 95 percent under the scenarios of population density decreasing 13, 39, and 47 percent, respectively.

In an attempt to summarize the voluminous literature available on scenario planning, researchers conducted a meta-analysis of scenarios that considered the impact of land use decisions on transportation.[538] The results suggest that increasing average regional density by 50 percent, directing development to infill locations, mixing land uses, and coordinating transportation investments could reduce VMT 17 percent below current trends between 2007 and 2050.

---

[536] National Research Council of the National Academies, *Driving and the Built Environment* 2009
[537] Marshall 2008. Values were chosen based on data showing average urban population density declined by 13 percent from 1960 to 1990 and by 34 percent from 1990 to 2000.
[538] Bartholomew and Ewing 2008

4.4   Summary

Research has shown that development decisions have both direct and indirect effects on the environment and that growth can be accommodated in ways that better protect the environment and human health. Strategies that minimize negative environmental impacts include modifying *where* we build to direct development away from sensitive natural areas and onto infill, brownfield, and greyfield sites while locating jobs, homes, and services near transit. Strategies also include modifying *how* we build to focus on more compact, mixed-use development that uses green building techniques and makes walking and biking convenient and enjoyable.

Used in combination, these practices can significantly reduce impacts to habitat, ecosystems, and watersheds and can reduce vehicle travel and energy use, which in turn reduces emissions that cause local, regional, and global air quality concerns.

Chapter 5. Conclusion

Across the country, communities are concerned about the built environment not just for quality of life and economic reasons, but also because of the effect that development has on human health, environmental resources, and natural habitats. *Our Built and Natural Environments* reviews the scientific literature and demonstrates that the built environment can significantly affect ecological and human health. As residents and public officials have come to understand the relationships among land use, transportation, and the environment, they have begun to seek new ways to grow—ways that benefit the environment and that support the jobs, economic development, health, and quality of life that depend on the protection of air and water quality.

How and where we build affects the natural environment and human health in the following ways:

- **Habitat and Ecosystems**—Development uses land and modifies habitats and ecosystems. In many metropolitan areas, the pace of land development continues to far exceed the pace of population growth. Not only does development directly destroy areas of natural habitat, it can fragment habitat and lead to the invasion of non-native species that severely alter ecosystem function and reduce biodiversity. Development that reuses and repurposes already-developed land takes development pressure off sensitive and critical habitats such as wetlands and forests. It can preserve ecosystem integrity and create amenities for adjacent neighborhoods.

- **Water Quality**—Development affects water quality by changing the natural flow of water in a watershed, particularly by increasing impervious surfaces and channeling stormwater runoff. At least 850,000 acres of lakes, reservoirs, and ponds and 50,000 miles of rivers and streams are impaired by stormwater runoff. As communities nationwide strive to protect their water resources, both for natural habitat and for clean drinking water, understanding the impact of development on water quality is important. Impervious surfaces increase runoff volumes and speed, which change the physical form of our stream systems and increase pollution in our waterways. Water quality can be improved by minimizing impervious surfaces through more compact, mixed-use development and using green infrastructure to manage stormwater where it falls.

- **Air Quality**—Air quality is related to how and where we build because building practices affect indoor air pollution levels and the energy needed to power buildings, and development patterns affect travel behavior. Gasoline-powered vehicles are significant contributors to air pollution. Although technology has significantly reduced per car vehicle emissions, the approximate 250 percent increase in VMT since 1970 has offset potential gains. There is significant evidence that compact, mixed-use development focused around transit can reduce vehicle travel and air pollution from motor vehicles. Infill development, including redevelopment of brownfields, often provides better access to transit services, which would reduce vehicle travel compared with development on the edge of the metropolitan area. In addition, designing roads to accommodate walkers and bikers safely and comfortably can encourage people to travel short distances without a car.

- **Global Climate**—Like air quality, global climate is affected by how and where we build because those decisions influence the energy needed to build and operate buildings and the amount people travel. Combustion of motor vehicle fuel emits carbon dioxide, a greenhouse gas that helps trap heat in the atmosphere, and transportation is responsible for 27 percent of U.S. greenhouse gas emissions. Residential and commercial buildings are responsible for 18 and 17 percent, respectively. Many communities understand that global warming is a serious threat and are encouraging practices that reduce greenhouse gas emissions. Examples include providing more transportation choices, reducing the need to travel by car, and improving the energy efficiency of buildings.

- **Contamination and Risk in Communities**—Old abandoned land in urbanized areas, potentially contaminated with hazardous or toxic waste, poses risks to communities. Redeveloping brownfields and hazardous waste sites provides the opportunity to clean up contaminated sites, reducing threats to water quality and human health. Brownfield and hazardous waste site redevelopment also uses existing infrastructure, including roads and water and wastewater systems, more efficiently, and it protects open space by placing new development in previously developed areas rather than in undisturbed habitat.

- **Public Health**—Where and how we build our communities affects not only the amount of pollution people are exposed to, but also how likely they are to get adequate physical exercise, feel a sense of well-being, and even suffer injury or death from a car crash. Designing communities so that walking and biking are convenient, practical, safe, and enjoyable for meeting daily needs can help achieve all of these aims.

Carefully choosing where and how we build can reduce the direct impacts of development on habitat, ecosystems, and water quality. Land use practices can also reduce indirect impacts on air quality and global climate by affecting travel choices. As communities nationwide look for ways to reduce the environmental and human health impacts of their development decisions, the evidence is clear that our nation can continue to grow and can build a strong foundation for lasting prosperity while also protecting our environment and health.

Works Cited

Adamkiewicz, Gary, et al. "Moving environmental justice indoors: Understanding structural influences on residential exposure patterns in low-income communities." *American Journal of Public Health* 101, no. S1 (2011): S238-S245.

Aguilar, Ramiro, Mauricio Quesada, Lorena Ashworth, Yvonne Herrerias-Diego, and Jorge Lobo. "Genetic consequences of habitat fragmentation in plant populations: Susceptible signals in plant traits and methodological approaches." *Molecular Ecology* 17, no. 24 (2008): 5177-5188.

Air Force Institute for Environment, Safety, and Occupational Health Risk Analysis. "Fact sheet: Indoor air and risk assessment." n.d. http://dhl.dhhq.health.mil/Product/RetrieveFile?prodId=32.

Al-Zoughool, Mustafa, and Daniel Krewski. "Health effects of radon: A review of the literature." *International Journal of Radiation Biology* 85, no. 1 (2009): 57-69.

Anderson, Brooke G., and Michelle L. Belle. "Weather-related mortality: How heat, cold, and heat waves affect mortality in the United States." *Epidemiology* 20 (2009): 205-213.

Arrington, G.B., and Robert Cervero. "Effects of TOD on housing, parking, and travel." *Transit Cooperative Research Program Report 128*. Transportation Research Board of the National Academies. 2008.

Bae, Chang-Hee Christine, Gail Sandlin, Alon Bassok, and Sungyop Kim. "The exposure of disadvantaged populations in freeway air-pollution sheds: A case study of the Seattle and Portland regions." *Environment and Planning B: Planning and Design* 34 (2007): 154-170.

Banks, Sam C., Maxine P. Piggott, Adam J. Stow, and Andrea C. Taylor. "Sex and sociality in a disconnected world: A review of the impacts of habitat fragmentation on animal social interactions." *Canadian Journal of Zoology* 85, no. 10 (2007): 1065-1079.

Bartholomew, Keith. "Land use-transportation scenario planning: Promise and reality." *Transportation* 34 (2007): 397-412.

Bartholomew, Keith, and Reid Ewing. "Land use-transportation scenarios and future vehicle travel and land consumption: A meta-analysis." *Journal of the American Planning Association* 75, no. 1 (2008): 13-27.

Basu, Rupa. "High ambient temperature and mortality: A review of epidemiological studies from 2001 to 2008." *Environmental Health* 8, no. 40 (2009).

Beal, Cara D., Edoardo Bertone, and Rodney A. Stewart. "Evaluating the energy and carbon reductions resulting from resource-efficient household stock." *Energy and Buildings* 55 (2012): 422-432.

Bean, Eban Zachary, William Frederick Hunt, and David Alan Bidelspach. "Evaluation of four permeable pavement sites in eastern North Carolina for runoff reduction and water quality impacts." *Journal of Irrigation and Drainage Engineering* 133, no. 6 (2007): 583-592.

Beaulieu, Karen M., Amanda H. Bell, and James F. Coles. *Variability in Stream Chemistry in Relation to Urban Development and Biological Condition in Seven Metropolitan Areas of the United States, 1999-2004*. U.S. Geological Survey. 2012. http://pubs.usgs.gov/sir/2012/5170/pdf/sir2012-5170_beaulieu_508.pdf.

Bechle, Matthew J., Dylan B. Millet, and Julian D. Marshall. "Effects of income and urban form on urban NO_2: Global evidence from satellites." *Environmental Science & Technology* 45 (2011): 4914-4919.

Bengston, David N., Jennifer O. Fletcher, and Kristen C. Nelson. "Public policies for managing urban growth and protecting open space: Policy instruments and lessons learned in the United States." *Landscape and Urban Planning* 69, no. 2-3 (2004): 271-286.

Bernstein, Jonathan A., et al. "The health effects of nonindustrial indoor air pollution." *Journal of Allergy and Clinical Immunology* 121, no. 3 (2008): 585-591.

Besser, Lilah M., and Andrew L. Dannenberg. "Walking to public transit: Steps to help meet physical activity recommendations." *American Journal of Preventive Medicine* 29, no. 4 (2005): 273-280.

Bhat, Chandra R., and Jessica Y. Guo. "A comprehensive analysis of built environment characteristics on household residential choice and auto ownership levels." *Transportation Research Part B: Methodological* 41, no. 5 (2007): 506-526.

Booth, Derek B., and Brian P. Bledsoe. "Streams and urbanization." In *The Water Environment of Cities*, by L. A. Baker, 93-123. New York: Springer, 2009.

Brattebo, Benjamin O., and Derek B. Booth. "Long-term stormwater quantity and quality performance of permeable pavement systems." *Water Research* 37, no. 18 (2003): 4369-4376.

Bribian, Ignacio Zabalza, Antonio Valero Capilla, and Alfonso Aranda Uson. "Life cycle assessment of building materials: Comparative analysis of energy and environmental impacts and evaluation of the eco-efficiency improvement potential." *Building and Environment* 46, no. 5 (2011): 1133-1140.

Brown, Barbara B., and Carol M. Werner. "A new rail stop: Tracking moderate physical activity bouts and ridership." *American Journal of Preventive Medicine* 33, no. 4 (2007): 306-309.

Brown, Thomas C., and Pamela Froemke. "Nationwide assessment of nonpoint source threats to water quality." *BioScience* 62, no. 2 (2012): 136-146.

Buehler, Ralph, and John Pucher. "Cycling to work in 90 large American cities: New evidence on the role of bike paths and lanes." *Transportation* 39 (2012): 409-432.

Burghardt, Karin T., Douglas W. Tallamy, and Gregory Shriver. "Impact of native plants on bird and butterfly biodiversity in suburban landscapes." *Conservation Biology* 23, no. 1 (2008): 219-224.

Burghardt, Karin T., Douglas W. Tallamy, Christopher Philips, and Kimberley J. Shropshire. "Non-native plants reduce abundance, richness, and host specialization in lepidopteran communities." *Ecosphere* 1, no. 5 (2010): 1-22.

Cao, Xinyu, Patricia L. Mokhtarian, and Susan L. Handy. "Examining the impacts of residential self-selection on travel behavior: A focus on empirical findings." *Transport Reviews* 29, no. 3 (2009): 359-395.

Carey, Richard O., et al. "Evaluating nutrient impacts in urban watersheds: Challenges and research opportunities." *Environmental Pollution* 173 (2013): 138-149.

Carlisle, Daren M., David M. Wolock, and Michael R. Meador. "Alteration of streamflow magnitudes and potential ecological consequences: A multiregional assessment." *Frontiers in Ecology and the Environment* 9, no. 5 (2011): 264-270.

Centers for Disease Control and Prevention, National Center for Injury Prevention and Control. *WISQARS Fatal Injury Reports, National and Regional, 1999-2010.* http://webappa.cdc.gov/sasweb/ncipc/mortrate10_us.html (accessed February 11, 2013).

—. *WISQARS Nonfatal Injury Reports.* http://webappa.cdc.gov/sasweb/ncipc/nfirates2001.html (accessed February 11, 2013).

Cervero, Robert. "Office development, rail transit, and commuting choice." *Journal of Public Transportation* 9, no. 5 (2006): 41-55.

—. "Road expansion, urban growth, and induced travel: A path analysis." *Journal of the American Planning Association* 69, no. 2 (2003): 145-163.

—. "Transit-oriented development's ridership bonus: A product of self-selection and public policies." *Environment and Planning A* 39 (2007): 2068-2085.

Cervero, Robert, and Jin Murakami. "Effects of built environments on vehicle miles traveled: Evidence from 370 U.S. urbanized areas." *Environment and Planning A* 42 (2010): 400-418.

Cervero, Robert, and Michael Duncan. "Which reduces vehicle travel more: Jobs-housing balance or retail-housing mixing?" *Journal of the American Planning Association* 72, no. 4 (2006): 475-490.

Chatman, Daniel G. "Residential choice, the built environment, and nonwork travel: Evidence using new data and methods." *Environment and Planning A* 41 (2009): 1072-1089.

Chattopadhyay, Sudip, and Emily Taylor. "Do smart growth strategies have a role in curbing vehicle miles traveled? A further assessment using household level survey data." *The B.E. Journal of Economic Analysis & Policy* 12, no. 1 (2012).

Chen, Li, Cynthia Chen, Reid Ewing, Claire E. McKnight, Raghavan Srinivasan, and Matthew Roe. "Safety countermeasures and crash reduction in New York City—Experience and lessons learned." *Accident Analysis and Prevention* 50 (2013): 312-322.

Chester, Mikhail, Arpad Horvath, and Samer Madanat. "Parking infrastructure: energy, emissions, and automobile life-cycle environmental accounting." *Environmental Research Letters* 5 (2010).

Clark, Lara P., Dylan B. Millet, and Julian D. Marshall. "Air quality and urban form in U.S. urban areas: Evidence from regulatory monitors." *Environmental Science & Technology* 45, no. 16 (2011): 7028-7035.

Cochran, K.M., and T.G. Townsend. "Estimating construction and demolition debris generation using a materials flow analysis approach." *Waste Management* 30, no. 11 (2010): 2247-2254.

Cochran, Kimberly, Timothy Townsend, Debra Reinhart, and Howell Heck. "Estimation of regional building-related C&D debris generation and composition: Case study for Florida, U.S." *Waste Management* 27, no. 7 (2007): 921-931.

Coffin, Alisa W. "From roadkill to road ecology: A review of the ecological effects of roads." *Journal of Transport Geography* 15, no. 5 (2007): 396-406.

Colbeck, Ian, and Zaheer Ahmad Nasir. "Indoor air pollution." In *Human Exposure to Pollutants via Dermal Absorption and Inhalation*, by Mihalis Lazaridis and Ian Colbeck, 41-72. Springer, 2010.

Coles, James F., Thomas F. Cuffney, Gerard McMahon, and Cornell J. Rosiu. "Judging a brook by its cover: The relation between ecological condition of a stream and urban land cover in New England." *Northeastern Naturalist* 17, no. 1 (2010): 29-48.

Costanza, Robert, Octavio Perez-Maqueo, Luisa Martinez, Paul Sutton, Sharolyn J. Anderson, and Kenneth Mulder. "The value of coastal wetlands for hurricane protection." *A Journal of the Human Environment* 37, no. 4 (2008): 241-248.

Crain, Caitlin M., Benjamin S. Halpern, Mike W. Beck, and Carrie V. Kappel. "Understanding and managing human threats to the coastal marine environment." *The Year in Ecology and Conservation Biology* 1162 (2009): 39-62.

Cuffney, Thomas F., Robin A. Brightbill, Jason T. May, and Ian R. Waite. "Responses of benthic macroinvertebrates to environmental changes associated with urbanization in nine metropolitan areas." *Ecological Applications* 20, no. 5 (2010): 1384-1401.

Cutts, Bethany B., Kate J. Darby, Christopher G. Boone, and Alexandra Brewis. "City structure, obesity, and environmental justice: An integrated analysis of physical and social barriers to walkable streets and park access." *Social Science & Medicine* 69 (2009): 1314-1322.

Dahl, T.E. *Status and Trends of Wetlands in the Conterminous United States 2004 to 2009.* U.S. Fish and Wildlife Service. 2011. http://www.fws.gov/wetlands/Status-And-Trends-2009/index.html.

Dahl, Thomas E. *Wetlands Losses in the United States 1780s to 1980s.* U.S. Fish and Wildlife Service. 1990.

Davis, Amelie Y., Bryan C. Pijanowski, Kimberly Robinson, and Bernard Engel. "The environmental and economic costs of sprawling parking lots in the United States." *Land Use Policy* 27 (2010): 255-261.

Davis, Mark A., et al. "Don't judge species on their origins." *Nature* 474 (2011): 153-154.

Davis, Stacy C., Susan W. Diegel, and Robert G. Boundy. *Transportation Energy Data Book: Edition 31.* U.S. Department of Energy. 2012. http://cta.ornl.gov/data/download31.shtml.

de Groot, Rudolf S., Matthew A. Wilson, and Roelof M.J. Boumans. "A typology for the classification, description and valuation of ecosystem functions, goods and services." *Ecological Economics* 41 (2002): 393-408.

Defeo, Omar, et al. "Threats to sandy beach ecosystems: A review." *Estuarine, Coastal and Shelf Science* 81 (2009): 1-12.

Di Giulio, Manuela, Rolf Hoderegger, and Silvia Tobias. "Effects of habitat and landscape fragmentation on humans and biodiversity in densely populated landscapes." *Journal of Environmental Management* 90, no. 10 (2009): 2959-2968.

DiCecio, Riccardo, Kristie M. Engemann, Michael T. Owyang, and Christopher H. Wheeler. "Changing trends in the labor force: A survey." *Federal Reserve Bank of St. Louis Review* 90, no. 1 (2008): 47-62.

Dietz, Michael E. "Low impact development practices: A review of current research and recommendations for future directions." *Water, Air, and Soil Pollution* 186, no. 1-4 (2007): 351-363.

Dill, Jennifer. "Transit use at transit-oriented developments in Portland, Oregon, area." *Transportation Research Record* 2063 (2008): 159-167.

Dodson, Robin E., et al. "After the PBDE phase-out: A broad suite of flame retardants in repeat house dust samples from California." *Environmental Science & Technology* 46 (2012): 13056-13066.

Dolan, Rebecca W., Marcia E. Moore, and Jessica D. Stephens. "Documenting effects of urbanization on flora using herbarium records." *Journal of Ecology* 99, no. 4 (2011): 1055-1062.

Drummond, Mark A., and Thomas R. Loveland. "Land-use pressure and a transition to forest-cover loss in the Eastern United States." *BioScience* 60, no. 4 (2010): 286-298.

D'Souza, Jennifer C., Chunrong Jia, Bhrarmar Mukherjee, and Stuart Batterman. "Ethnicity, housing and personal factors as determinants of VOC exposures." *Atmospheric Environment* 18, no. 43 (2009): 2884-2892.

Duke, J., M. Huhman, and C. Heitzler. "Physical activity levels among children aged 9-13 years—United States, 2002." *Morbidity and Mortality Weekly Report* 52, no. 33 (2003): 785-788.

Dulal, Hari Bansha, Gernot Brodnig, and Charity G. Onoriose. "Climate change mitigation in the transport sector through urban planning: A review." *Habitat International* 35, no. 3 (2011): 494-500.

Dumbaugh, Eric, and Robert Rae. "Safe urban form: Revisiting the relationship between community design and traffic safety." *Journal of the American Planning Association* 75, no. 3 (2009): 309-329.

Dumbaugh, Eric, and Wenhao Li. "Designing for the safety of pedestrians, cyclists, and motorists in urban environments." *Journal of the American Planning Association* 77, no. 1 (2010): 69-88.

Duranton, Gilles, and Matthew A. Turner. "The fundamental law of road congestion: Evidence from U.S. cities." *American Economic Review* 101 (2011): 2616-2652.

Durkin, Thomas H. "Boiler system efficiency." *ASHRAE Journal* 48 (2006): 51-57.

EPA. "1970-2012 average annual emissions, all criteria pollutants in MS Excel." *National Emissions Inventory (NEI) Air Pollutant Emissions Trends Data.* 2012. http://www.epa.gov/ttn/chief/trends/index.html.

—. *2005 National-Scale Air Toxics Assessment.* http://www.epa.gov/ttn/atw/nata2005/index.html (accessed October 31, 2012).

—. *About Air Toxics.* http://www.epa.gov/ttn/atw/allabout.html (accessed October 31, 2012).

—. *About Energy Star.* http://www.energystar.gov/index.cfm?c=about.ab_index (accessed November 8, 2012).

—. *Air Emission Sources.* 2008. http://www.epa.gov/air/emissions/index.htm (accessed September 24, 2012).

—. "Air toxics pie charts." *2005 Assessment Results.* 2011. http://www.epa.gov/ttn/atw/nata2005/tables.html.

—. *The Ambient Air Monitoring Program.* http://epa.gov/air/oaqps/qa/monprog.html (accessed April 4, 2013).

—. *Cleaning up the Nation's Waste Sites: Markets and Technology Trends.* 2004. http://www.epa.gov/superfund/accomp/news/30years.htm.

—. *Design for the Environment.* http://www.epa.gov/dfe (accessed February 21, 2013).

—. *Distribution System Inventory, Integrity and Water Quality.* 2007. http://www.epa.gov/ogwdw/disinfection/tcr/pdfs/issuepaper_tcr_ds-inventory.pdf.

—. *Estimating 2003 Building-Related Construction and Demolition Materials Amounts.* 2009. http://www.epa.gov/wastes/conserve/imr/cdm/pubs/cd-meas.pdf.

—. *FY 2011-2015 EPA Strategic Plan: Achieving Our Vision.* 2010. http://www.epa.gov/planandbudget/strategicplan.html.

—. *Greenhouse Gas Emissions.* http://epa.gov/climatechange/ghgemissions (accessed September 26, 2012).

—. *Health Effects of PCBs.* http://www.epa.gov/epawaste/hazard/tsd/pcbs/pubs/effects.htm (accessed February 11, 2013).

—. *Inventory of U.S. Greenhouse Gas Emissions and Sinks: 1990-2010.* 2012. http://www.epa.gov/climatechange/Downloads/ghgemissions/US-GHG-Inventory-2012-Main-Text.pdf.

—. *Learn About Asbestos.* http://www2.epa.gov/asbestos/learn-about-asbestos (accessed February 11, 2013).

—. *National Ambient Air Quality Standards (NAAQS).* http://www.epa.gov/air/criteria.html (accessed September 21, 2012).

—. *Opportunities to Reduce Greenhouse Gas Emissions through Materials and Land Management Practices.* 2009. http://www.epa.gov/oswer/docs/ghg_land_and_materials_management.pdf.

—. *Our Nation's Air: Status and Trends Through 2008.* 2010. http://www.epa.gov/airtrends/2010/report/fullreport.pdf.

—. *Our Nation's Air: Status and Trends Through 2010.* 2012. http://www.epa.gov/airtrends/2011/report/fullreport.pdf.

—. *Protecting Water Resources with Higher-Density Development.* 2006. http://www.epa.gov/smartgrowth/water_density.htm.

—. *Residential Construction Trends in America's Metropolitan Regions.* 2012. http://www.epa.gov/smartgrowth/construction_trends.htm.

—. *Summary Nonattainment Area Population Exposure Report.* 2012. http://www.epa.gov/airquality/greenbk/popexp.html.

—. *Technical Fact Sheet – Polybrominated Diphenyl Ethers (PBDEs) and Polybrominated Biphenyls (PBBs).* 2012. http://www.epa.gov/fedfac/pdf/technical_fact_sheet_pbde_pbb.pdf.

—. *WaterSense.* http://www.epa.gov/WaterSense/about_us/index.html (accessed February 26, 2013).

—. *WaterSense at Work: Best Management Practices for Commercial and Institutional Facilities.* 2012. http://www.epa.gov/watersense/commercial/docs/watersense_at_work/#/1.

—. *Watershed Assessment, Tracking & Environmental ResultS.* http://ofmpub.epa.gov/tmdl_waters10/attains_nation_cy.control (accessed September 10, 2012).

—. *What is Green Infrastructure?* http://water.epa.gov/infrastructure/greeninfrastructure/gi_what.cfm (accessed February 13, 2013).

Evans, John E. *Traveler Response to Transportation System Changes. Chapter 9—Transit Scheduling and Frequency.* Transportation Research Board of the National Academies. 2004. http://onlinepubs.trb.org/onlinepubs/tcrp/tcrp_rpt_95c9.pdf.

Ewing, Reid. "Highway-induced development: Research results for metropolitan areas." *Transportation Research Record: Journal of the Transportation Research Board* 2067 (2008): 101-109.

Ewing, Reid, and Eric Dumbaugh. "The built environment and traffic safety: A review of empirical evidence." *Journal of Planning Literature* 23, no. 4 (2009): 347-367.

Ewing, Reid, and Robert Cervero. "Travel and the built environment: A meta-analysis." *Journal of the American Planning Association* 76, no. 3 (2010): 265-294.

Ewing, Reid, et al. "Traffic generated by mixed-use developments—Six-region study using consistent built environment measures." *Journal of Urban Planning and Development* 137, no. 3 (2011): 248-261.

Ewing, Reid, Keith Bartholomew, Steve Winkelman, Jerry Walters, and Geoffrey Anderson. "Urban development and climate change." *Journal of Urbanism* 1, no. 3 (2008): 201-216.

Ewing, Reid, Richard A. Schieber, and Charles V. Zegeer. "Urban sprawl as a risk factor in motor vehicle occupant and pedestrian fatalities." *American Journal of Public Health* 93, no. 9 (2003): 1541-1545.

Fan, Yingling, and Asad J. Khattak. "Urban form, individual spatial footprints, and travel: Examination of space-use behavior." *Transportation Research Record* 2082 (2008): 98-106.

Fann, Neal, Amy D. Lamson, Susan C. Anenberg, Karen Wesson, David Risley, and Bryan J. Hubbell. "Estimating the national public health burden associated with exposure to ambient $PM_{2.5}$ and ozone." *Risk Analysis* 32, no. 1 (2012): 81-95.

Faulkner, Guy E.J., Ron N. Buliung, Parminder K. Flora, and Caroline Fusco. "Active school transport, physical activity levels and body weight of children and youth: A systematic review." *Preventive Medicine* 48, no. 1 (2009): 3-8.

Federal Highway Administration. *Data Extraction and Visualization Prototypes.* http://nhts.ornl.gov/det (accessed January 22, 2013).

—. *Highway Statistics.* 2010. http://www.fhwa.dot.gov/policyinformation/statistics/2010.

—. *Historical Monthly VMT Report.* March 26, 2012.

—. *National Household Travel Survey: Our Nation's Travel.* http://nhts.ornl.gov (accessed January 22, 2013).

—. *Summary of Travel Trends: 2009 National Household Travel Survey.* 2011. http://nhts.ornl.gov/2009/pub/stt.pdf.

—. *Wildlife-Vehicle Collision Reduction Study: Report To Congress.* 2008. http://www.fhwa.dot.gov/publications/research/safety/08034/index.cfm.

Federal Interagency Stream Restoration Working Group. *Stream Corridor Restoration: Principles, Processes, and Practices.* 1998.

Feng, Jing, Thomas A. Glass, Frank C. Curriero, Walter F. Steward, and Brian A. Schwartz. "The built environment and obesity: A systematic review of the epidemiologic evidence." *Health & Place* 16 (2010): 175-190.

Ferdinand, Alva O., Bisakha Sen, Saurabh Rahurkar, Sally Engler, and Nir Menachemi. "The relationship between built environments and physical activity: A systematic review." *American Journal of Public Health* 102, no. 10 (2012): e7-e13.

Fischer, Joern, and David B. Lindenmayer. "Landscape modification and habitat fragmentation: A synthesis." *Global Ecology and Biogeography* 16, no. 3 (2007): 265-280.

Fitzpatrick, Faith A., and Marie C. Peppler. *Relation of Urbanization to Stream Habitat and Geomorphic Characteristics in Nine Metropolitan Areas of the United States.* U.S. Geological Survey. 2010. http://pubs.usgs.gov/sir/2010/5056.

Flegal, Katherine M., Margaret D. Carroll, Cynthia L. Ogden, and Lester R. Curtin. "Prevalence and trends in obesity among U.S. adults, 1999-2008." *The Journal of the American Medical Association* 303, no. 3 (2010): 235-241.

Forman, Richard T.T. "Estimate of the area affected ecologically by the road system in the United States." *Conservation Biology* 14, no. 1 (2000): 31-35.

Forman, Richard T.T., and Robert D. Deblinger. "The ecological road-effect zone of a Massachusetts (U.S.A.) suburban highway." *Conservation Biology* 14, no. 1 (2000): 36-46.

Forman, Richard T.T., et al. *Road Ecology: Science and Solutions.* Washington: Island Press, 2003.

Frank, Lawrence D., and Peter Engelke. "Multiple impacts of the built environment on public health: Walkable places and the exposure to air pollution." *International Regional Science Review* 28, no. 2 (2005): 193-216.

Frank, Lawrence D., James F. Sallis, Terry L. Conway, James E. Chapman, and Brian E. Saelens. "Many pathways from land use to health: Associations between neighborhood walkability and active transportation, body mass index, and air quality." *Journal of the American Planning Association* 72, no. 1 (2006): 75-87.

Frazer, Lance. "Paving paradise: The peril of impervious surfaces." *Environmental Health Perspectives* 113, no. 7 (2005): A456-A462.

Frumkin, Howard. "Cities, suburbs, and urban sprawl: Their impact on health." In *Cities and the Health of the Public*, by Nicholas Freudenberg and Sandro Galea, 143-175. Vanderbilt University Press, 2006.

Frumkin, Howard. "Urban sprawl and public health." *Public Health Reports* 117 (2002): 201-217.

Fuentes-Leonarte, Virginia, Jose M. Tenias, and Ferran Ballester. "Levels of pollutants in indoor air and respiratory health in preschool children: A systematic review." *Pediatric Pulmonology* 44 (2009): 231-243.

Gebel, Klaus, Adrian E. Bauman, and Mark Petticrew. "The physical environment and physical activity: A critical appraisal of review articles." *American Journal of Preventive Medicine* 32, no. 5 (2007): 361-369.

Giles, Luisa V., et al. "From good intentions to proven interventions: Effectiveness of actions to reduce the health impacts of air pollution." *Environmental Health Perspectives* 199, no. 1 (2011): 29-36.

Gilliom, Robert J., et al. *Pesticides in the Nation's Streams and Ground Water 1992-2001.* U.S. Geological Survey. 2006. http://pubs.usgs.gov/circ/2005/1291/pdf/circ1291.pdf.

Gim, Tae-Hyoung Tommy. "A meta-analysis of the relationship between density and travel behavior." *Transportation* 39 (2012): 491-519.

Gordon-Larson, Penny, Melissa C. Nelson, Phil Page, and Barry M. Popkin. "Inequality in the built environment underlies key health disparities in physical activity and obesity." *Pediatrics* 117, no. 2 (2006): 417-424.

Gregory, Justin H., Michael D. Dukes, Pierce H. Jones, and Grady L. Miller. "Effect of urban soil compaction on infiltration rate." *Journal of Soil and Water Conservation* 61, no. 3 (2006): 117-124.

Haley, Melissa B., Michael D. Dukes, and Grady L. Miller. "Residential irrigation water use in central Florida." *Journal of Irrigation and Drainage Engineering* 133, no. 5 (2007): 427-434.

Handy, Susan. "Smart growth and the transportation-land use connection: What does the research tell us?" *International Regional Science Review* 28, no. 2 (2005): 146-167.

Harvey, L.D. Danny. "Reducing energy use in the buildings sector: Measures, costs, and examples." *Energy Efficiency* 2 (2009): 139-163.

Hassan, Rashid, Robert Scholes, and Neville Ash. *Ecosystems and Human Well-being: Current State and Trends, Volume 1.* Island Press. 2005. http://www.unep.org/maweb/en/Condition.aspx.

Hatt, Belinda E., Tim D. Fletcher, Christopher J. Walsh, and Sally I. Taylor. "The influence of urban density and drainage infrastructure on the concentrations and loads of pollutants in small streams." *Environmental Management* 34, no. 1 (2004): 112-124.

Heath, Gregory W., Ross C. Brownson, Judy Kruger, Rebecca Miles, Kenneth E. Powell, and Leigh T. Ramsey. "The effectiveness of urban design and land use and transport policies and practices to increase physical activity: A systematic review." *Journal of Physical Activity and Health* 3, no. Suppl 1 (2006): S55-S76.

Hickman, Jonathan E., Shiliang Wu, Loretta J. Mickley, and Manuel T. Lerdau. "Kudzu (Pueraria montana) invasion doubles emissions of nitric oxide and increases ozone pollution." *Proceedings of the National Academy of Sciences* 107, no. 22 (2010): 10115-10119.

Horvath, Arpad. "Construction materials and the environment." *Annual Review of Environment and Resources* 29 (2004): 181-204.

Hostetler, Mark E., and Martin B. Main. "Native landscaping vs. exotic landscaping: What should we recommend?" *Journal of Extension* 45, no. 5 (2010).

Hun, Diana E., Jeffrey A. Siegel, Maria T. Morandi, Thomas H. Stock, and Richard L. Corsi. "Cancer risk disparities between Hispanic and non-Hispanic white populations: The role of exposure to indoor air pollution." *Environmental Health Perspectives* 117, no. 12 (2009): 1925-1931.

Hurd, Brian H. "Water conservation and residential landscapes: Household preferences, household choices." *Journal of Agricultural and Resource Economics* 31, no. 2 (2006): 173-192.

Intergovernmental Panel on Climate Change. *Climate Change 2007: Synthesis Report.* 2007. http://www.ipcc.ch/publications_and_data/publications_ipcc_fourth_assessment_report_synthesis_report.htm.

Jackson, Laura E. "The relationship of urban design to human health and condition." *Landscape and Urban Planning* 64 (2003): 191-200.

Jacob, John S., and Ricardo Lopez. "Is denser greener? An evaluation of higher density development as an urban stormwater-quality best management practice." *Journal of the American Water Resources Association* 45, no. 3 (2009): 687-701.

Jacobson, Carol R. "Identification and quantification of the hydrological impacts of imperviousness in urban catchments: A review." *Journal of Environmental Management* 92, no. 6 (2011): 1438-1448.

James, J.P., and X. Yang. "Emissions of volatile organic compounds from several green and non-green building materials: A comparison." *Indoor and Built Environment* 14, no. 1 (2004): 69-74.

Johnson, Paula I., Heather M. Stapleton, Andreas Sjodin, and John D. Meeker. "Relationships between polybrominated diphenyl ether and concentrations in house dust and serum." *Environmental Science & Technology* 44, no. 14 (2010): 5627-5632.

Jun, Myung-Jin. "Are Portland's smart growth policies related to reduced automobile dependence?" *Journal of Planning Education and Research* 28 (2008): 100-107.

Kahyaoglu-Koracin, Julide, Scott D. Bassett, David A. Mouat, and Alan W. Gertler. "Application of a scenario-based modeling system to evaluate the air quality impacts of future growth." *Atmospheric Environment* 43, no. 5 (2009): 1021-1028.

Keddy, Paul A. *Wetland Ecology: Principles and Conservation.* 2nd. Cambridge, UK: Cambridge University Press, 2010.

Kenny, Joan F., Nancy L. Barber, Susan S. Hutson, Kristin S. Linsey, John K. Lovelace, and Molly A. Maupin. *Estimated Use of Water in the United States in 2005.* U.S. Geological Survey. 2009. http://pubs.usgs.gov/circ/1344.

King, Ryan S., Matthew E. Baker, Paul F. Kazyak, and Donald E. Weller. "How novel is too novel? Stream community thresholds at exceptionally low levels of catchment urbanization." *Ecological Applications* 21, no. 5 (2011): 1659-1678.

Kockelman, Kara, Matthew Bomberg, Melissa Thompson, and Charlotte Whitehead. *GHG emissions control options: Opportunities for conservation.* Commissioned for *Driving and the Built Environment: The Effects of Compact Development on Motorized Travel, Energy Use, and CO$_2$ Emissions—Special Report 298.* National Academy of Sciences. 2009. http://onlinepubs.trb.org/Onlinepubs/sr/sr298kockelman.pdf.

Krizek, Kevin J. "Residential relocation and changes in urban travel: Does neighborhood-scale urban form matter?" *Journal of the American Planning Association* 69, no. 3 (2003): 265-281.

Kuzmyak, J. Richard, Richard H. Pratt, G. Bruce Douglas, and Frank Spielberg. *Traveler Response to Transportation System Changes: Chapter 15—Land Use and Site Design.* Transportation Research Board. 2003. http://onlinepubs.trb.org/onlinepubs/tcrp/tcrp_rpt_95c15.pdf.

Landis, John D., Heather Hood, Guangyu Li, Thomas Rogers, and Charles Warren. "The future of infill housing in California: Opportunities, potential, and feasibility." *Housing Policy Debate* 17, no. 4 (2006): 681-726.

Lavoie, Emma T., et al. "Chemical alternatives assessment: Enabling substitution to safer chemicals." *Environmental Science & Technology* 44, no. 24 (2010): 9244-9249.

Lee, Mengshan, and Berrin Tansel. "Life cycle based analysis of demands and emissions for residential water-using appliances." *Journal of Environmental Management* 101 (2012): 75-81.

Lee, Sangyun, and Paul Mohai. "Racial and socioeconomic assessments of neighborhoods adjacent to small-scale brownfield sites in the Detroit region." *Environmental Practice* 13 (2011): 340-353.

Leigh, Nancey Green, and Sarah L. Coffin. "Modeling the relationship among brownfields, property values, and community revitalization." *Housing Policy Debate* 16, no. 2 (2005): 257-280.

Lerman, Susannah B., and Paige S. Warren. "The conservation value of residential yards: Linking birds and people." *Ecological Applications* 21, no. 4 (2011): 1327-1339.

Levine, Jonathan, and Lawrence D. Frank. "Transportation and land-use preferences and residents' neighborhood choices: The sufficiency of compact development in the Atlanta region." *Transportation* 34 (2007): 255-274.

Levine, Jonathan, Aseem Inam, and Gwo-Wei Torng. "A choice-based rationale for land use and transportation alternatives: Evidence from Boston and Atlanta." *Journal of Planning Education and Research* 24 (2005): 317-330.

Lewis, Paul, William Rasdorf, H. Christopher Frey, Shih-Hao Pang, and Kangwook Kim. "Requirements and incentives for reducing construction vehicle emissions and comparison of nonroad diesel engine emissions data sources." *Journal of Construction Engineering and Management* 135, no. 5 (2009): 341-351.

Li, Fuzhong, et al. "Built environment, adiposity, and physical activity in adults aged 50-75." *American Journal of Preventive Medicine* 35, no. 1 (2008): 38-46.

Line, D.E., and N.M. White. "Effects of development on runoff and pollutant export." *Water Environment Research* 79, no. 2 (2007): 185-190.

LoGiudice, Kathleen, Richard S. Ostfeld, Kenneth A. Schmidt, and Felicia Keesing. "The ecology of infectious disease: Effects of host diversity and community composition on Lyme disease risk." *Proceedings of the National Academy of Sciences* 100, no. 2 (2003): 567-571.

Lubans, David R., Colin A. Boreham, Paul Kelly, and Charlie E. Foster. "The relationship between active travel to school and health-related fitness in children and adolescents: A systematic review." *International Journal of Behavioral Nutrition and Physical Activity* 8, no. 5 (2011): 1-12.

Lubeck, Aaron, and Francis Conlin. "Efficiency and comfort through deep energy retrofits: Balancing energy and moisture management." *Journal of Green Building* 3, no. 5 (2010).

MacDonald, John M., Robert J. Stokes, Deborah A. Cohen, Aaron Kofner, and Gred K. Ridgeway. "The effect of light rail transit on body mass index and physical activity." *American Journal of Preventive Medicine* 39, no. 2 (2010): 105-112.

Mahmoud, Mohammed, et al. "A formal framework for scenario development in support of environmental decision-making." *Environmental Modelling & Software* 24, no. 7 (2009): 798-808.

Manville, Michael, and Donald Shoup. "Parking, people, and cities." *Journal of American Planning and Development* 131, no. 4 (2005): 233-245.

Marshall, Julian D. "Energy-efficient urban form: Reducing urban sprawl could play an important role in addressing climate change." *Environmental Science & Technology* 42, no. 9 (2008): 3133-3137.

Marshall, Julian D., Michael Brauer, and Lawrence D. Frank. "Healthy neighborhoods: Walkability and air pollution." *Environmental Health Perspectives* 117, no. 11 (2009): 1752-1759.

Marshall, Julian D., Thomas E. McKone, Elizabeth Deakin, and William W. Nazaroff. "Inhalation of motor vehicle emissions: Effects of urban population and land area." *Atmospheric Environment* 39, no. 2 (2005): 283-295.

Massada, Avi Bar, Volker C. Radeloff, Susan I. Stewart, and Todd J. Hawbaker. "Wildfire risk in the wildland-urban interface: A simulation study in northwestern Wisconsin." *Forest Ecology and Management* 258, no. 9 (2009): 1990-1999.

McCarthy, Linda. "Efficiency in targeting the most marketable sites rather than equity in public assistance for brownfield redevelopment." *Economic Development Quarterly* 23, no. 3 (2009): 211-228.

McConville, Megan E., Daniel A. Rodriguez, Kelly Clifton, Gihyoug Cho, and Sheila Fleischhacker. "Disaggregate land uses and walking." *American Journal of Preventive Medicine* 40, no. 1 (2011): 25-32.

McDonald, Noreen C., Austin L. Brown, Lauren M. Marchetti, and Margo S. Pedroso. "U.S. school travel, 2009: An assessment of trends." *American Journal of Preventive Medicine*, 2011: 146-151.

McKinney, Michael L. "Urbanization as a major cause of biotic homogenization." *Biological Conservation* 127 (2006): 247-260.

Memmott, Jeffery. *Trends in Personal Income and Passenger Vehicle Miles.* Bureau of Transportation Statistics. 2007.

Mendell, M.J. "Indoor residential chemical emissions as risk factors for respiratory and allergic effects in children: A review." *Indoor Air* 17, no. 4 (2007): 259-277.

Mendell, Mark J., Anna G. Mirer, Kerry Cheung, My Tong, and Jeroen Douwes. "Respiratory and allergic health effects of dampness, mold, and dampness-related agents: A review of the epidemiologic evidence." *Environmental Health Perspectives* 119, no. 6 (2011): 748-756.

Metz, David. "The myth of travel time saving." *Transport Reviews* 28, no. 3 (2008): 321-336.

Millard-Ball, Adam, and Lee Schipper. "Are we reaching peak travel? Trends in passenger transport in eight industrialized countries." *Transport Reviews* 31, no. 3 (2011): 357-378.

Miranda, Marie Lynn, Sharon E. Edwards, Martha H. Keating, and Christopher J. Paul. "Making the environmental justice grade: The relative burden of air pollution exposure in the United States." *International Journal of Environmental Research and Public Health* 8 (2011): 1755-1771.

Mokhtarian, Patricia L., Franciso J. Samaniego, Robert H. Shumway, and Neil H. Willits. "Revisiting the notion of induced traffic through a matched-pairs study." *Transportation* 29 (2002): 193-220.

Moore, Latetia V., Ana V. Diez Roux, Kelly R. Evenson, Aileen P. McGinn, and Shannon J. Brines. "Availability of recreational resources in minority and low socioeconomic status areas." *American Journal of Preventive Medicine* 34, no. 1 (2008): 16-22.

Moran, Patrick W., et al. *Contaminants in Stream Sediments from Seven U.S. Metropolitan Areas: Data Summary of a National Pilot Study.* U.S. Geological Survey. 2012. http://pubs.usgs.gov/sir/2011/5092.

Muench, Stephen T. "Roadway construction sustainability impacts: Review of life-cycle assessments." *Transportation Research Record* 2151 (2010): 36-45.

Mukhija, Vinit, and Donald Shoup. "Quantity versus quality in off-street parking requirements." *Journal of the American Planning Association* 72, no. 3 (2006): 296-308.

Myers, Dowell, and John Pitkin. "Demographic forces and turning points in the American city, 1950-2040." *The Annals of the American Academy of Political and Social Science* 626, no. 1 (2009): 91-111.

National Highway Traffic Safety Administration. *CAFE – Fuel Economy.* http://www.nhtsa.gov/fuel-economy (accessed February 26, 2013).

—. *FARS Data Tables: Summary.* http://www-fars.nhtsa.dot.gov/Main/index.aspx (accessed October 4, 2012).

—. *Traffic Safety Facts 2000.* 2001. http://www-nrd.nhtsa.dot.gov/Pubs/TSF2000.pdf.

—. *Traffic Safety Facts 2010.* 2012. http://www-nrd.nhtsa.dot.gov/Pubs/811659.pdf.

National Invasive Species Council. *Welcome to InvasiveSpecies.gov.* http://invasivespecies.gov (accessed March 25, 2013).

National Oceanic and Atmospheric Administration. *Climate: U.S. Population in the Coastal Floodplain.* http://stateofthecoast.noaa.gov/pop100yr/welcome.html (accessed February 14, 2013).

—. *Climate: Vulnerability of our Nation's Coasts to Sea Level Rise.* http://stateofthecoast.noaa.gov/vulnerability/welcome.html (accessed February 14, 2013).

—. *National Coastal Population Report: Population Trends from 1970 to 2020.* 2013. http://stateofthecoast.noaa.gov/features/coastal-population-report.pdf.

National Research Council of the National Academies. *America's Climate Choices.* The National Academies Press. 2011. http://nas-sites.org/americasclimatechoices/sample-page/panel-reports/americas-climate-choices-final-report.

—. *Driving and the Built Environment: The Effects of Compact Development on Motorized Travel, Energy Use, and CO_2 Emissions—Special Report 298.* Transportation Research Board. 2009. http://www.nap.edu/catalog.php?record_id=12747.

—. *Urban Stormwater Management in the United States.* The National Academies Press. 2009. http://www.nap.edu/catalog.php?record_id=12465.

Nawaz, Muhammad Farrakh, Guilhem Bourrie, and Fabienne Trolard. "Soil compaction impact and modelling. A review." *Agronomy for Sustainable Development* 33, no. 2 (2013): 291-309.

Nelson, Arthur C. "The new urbanity: The rise of a new America." *The Annals of the American Academy of Political and Social Science* 626 (2009): 192-208.

Niemeier, Deb, Song Bai, and Susan Handy. "The impact of residential growth patterns on vehicle travel and pollutant emissions." *The Journal of Transport and Land Use* 4, no. 3 (2011): 65-80.

Noland, Robert B., and Lewison L. Lem. "A review of the evidence for induced travel and changes in transportation and environmental policy in the U.S. and the U.K." *Transportation Research Part D: Transport and Environment* 7, no. 1 (2002): 1-26.

Norman, Jonathan, Heather L. MacLean, and Christopher A. Kennedy. "Comparing high and low residential density: Life-cycle analysis of energy use and greenhouse gas emissions." *Journal of Urban Planning and Development* 132, no. 1 (2006): 10-21.

Nowak, David J., and Eric J. Greenfield. "Tree and impervious cover change in U.S. cities." *Urban Forestry & Urban Greening* 11 (2012): 21-30.

Nowak, David J., and Jeffrey T. Walton. "Projected urban growth (2000-2050) and its estimated impact on the U.S. forest resource." *Journal of Forestry* 103, no. 8 (2005): 383-389.

Ochoa, Luis, Chris Hendrickson, and Scott Matthews. "Economic input-output life-cycle assessment of U.S. residential buildings." *Journal of Infrastructure Systems* 8, no. 4 (2002): 132-138.

Ogden, Cynthia L., and Margaret D. Carroll. *Prevalence of Overweight, Obesity, and Extreme Obesity Among Adults: United States, Trends 1960-1962 Through 2007-2008.* National Center for Health Statistics. 2010.

Ory, David T., Patricia L. Mokhtarian, Lothlorien S. Redmond, Ilan Salomon, Gustavo O. Collantes, and Sangho Choo. "When is commuting desirable to the individual?" *Growth and Change* 35, no. 3 (2004): 334-359.

Page, G.W., and R.S. Berger. "Characteristics and land use of contaminated brownfield properties in voluntary cleanup agreement programs." *Land Use Policy* 23, no. 4 (2006): 551-559.

Perez-Padilla, R., A. Schilmann, and H. Riojas-Rodriguez. "Respiratory health effects of indoor air pollution." *The International Journal of Tuberculosis and Lung Disease* 14, no. 9 (2010): 1079-1086.

Pickett, S.T.A., et al. "Urban ecological systems: Scientific foundations and a decade of progress." *Journal of Environmental Management* 92, no. 3 (2011): 331-362.

Pimentel, David, Rodolfo Zuniga, and Doug Morrison. "Update on the environmental and economic costs associated with alien-invasive species in the United States." *Ecological Economics* 52, no. 3 (2005): 273-288.

Poff, N. LeRoy, Brian P. Bledsoe, and Christopher O. Cuhaciyan. "Hydrologic variation with land use across the contiguous United States: Geomorphic and ecological consequences for stream ecosystems." *Geomorphology* 79 (2006): 264-285.

Powell, Scott L., Warren B. Cohen, Zhiqiang Yang, John D. Pierce, and Marina Alberti. "Quantification of impervious surface in the Snohomish Water Resources Inventory Area of western Washington from 1972-2006." *Remote Sensing of Environment* 112, no. 4 (2008): 1895-1908.

Pratt, Richard H., and John E. Evans. *Traveler Response to Transportation System Changes. Chapter 10—Bus Routing and Coverage.* Transportation Research Board of the National Academies. 2004. http://trid.trb.org/view.aspx?id=705130.

Price, Katie. "Effects of watershed topography, soils, land use, and climate on baseflow hydrology in humid regions: A review." *Progress in Physical Geography* 35, no. 4 (2011): 465-492.

Pucher, John, Jennifer Dill, and Susan Handy. "Infrastructure, programs, and policies to increase bicycling: An international review." *Preventive Medicine* 50 (2010): S106-S125.

Pysek, Petr, and David M. Richardson. "Invasive species, environmental change and management, and health." *Annual Review of Environment and Resources* 35 (2010): 25-55.

Pysek, Petr, et al. "A global assessment of invasive plant impacts on resident species, communities and ecosystems: The interaction of impact measures, invading species' traits and environment." *Global Change Biology* 18 (2012): 1725-1737.

Radeloff, Volker C., et al. "Housing growth in and near United States protected areas limits their conservation value." *Proceedings of the National Academy of Sciences* 107, no. 2 (2010): 940-945.

Rahman, Naureen Mahbub, and Bliss L. Tracy. "Radon control systems in existing and new construction: A review." *Radiation Protection Dosimetry* 135, no. 4 (2009): 243-255.

Ramankutty, Navin, Elizabeth Heller, and Jeanine Rhemtulla. "Prevailing myths about agricultural abandonment and forest regrowth in the United States." *Annals of the Association of American Geographers* 100, no. 3 (2010): 502-512.

Redmond, Lothlorien S., and Patricia L. Mokhtarian. "The positive utility of the commute: Modeling ideal commute time and relative desired commute amount." *Transportation* 28, no. 2 (2001): 179-205.

Reynolds, Conor C.O., M. Anne Harris, Kay Teschke, Peter A. Cripton, and Meghan Winters. "The impact of transportation infrastructure on bicycling injuries and crashes: A review of the literature." *Environmental Health* 8 (2009): 1-19.

Riitters, Kurt H., and James D. Wickham. "How far to the nearest road?" *Frontiers in Ecology and the Environment* 1, no. 3 (2003): 125-129.

Rodier, Caroline. "Transit, land use, and auto pricing strategies to reduce vehicle miles traveled and greenhouse gas emissions." *Transportation Research Record: Journal of the Transportation Research Board* 2132 (2009): 1-12.

Rodriguez, Daniel A., Kelly R. Evenson, Ana V. Roux, and Shannon J. Brines. "Land use, residential density, and walking: The multi-ethnic study of atherosclerosis." *American Journal of Preventive Medicine* 37, no. 5 (2009): 397-404.

Rogers, Shannon H., John M. Halstead, Kevin H. Gardner, and Cynthia H. Carlson. "Examining walkability and social capital as indicators of quality of life at the municipal and neighborhood scales." *Applied Research in Quality of Life* 6 (2011): 201-213.

Rosen, Erik, Helena Stigson, and Ulrich Sander. "Literature review of pedestrian fatality risk as a function of car impact speed." *Accident Analysis and Prevention* 43 (2011): 25-33.

Rowe, D. Bradley. "Green roofs as a means of pollution abatement." *Environmental Pollution* 159, no. 8-9 (2011): 2100-2110.

Rudel, Ruthann A., and Laura J. Perovich. "Endocrine disrupting chemicals in indoor and outdoor air." *Atmospheric Environment* 43, no. 1 (2009): 170-181.

Saelens, Brian E., and Susan L. Handy. "Built environment correlates of walking: A review." *Medicine and Science in Sports and Exercise* 40, no. 7 (2008): S550-S566.

Sailor, David J., Timothy B. Elley, and Max Gibson. "Exploring the building energy impacts of green roof design decisions—A modeling study of buildings in four distinct climates." *Journal of Building Physics* 35, no. 4 (2012): 372-391.

Salon, Deborah, Marlon G. Boarnet, Susan Handy, Steven Spears, and Gil Tal. "How do local actions affect VMT? A critical review of the empirical evidence." *Transportation Research Part D* 17 (2012): 495-508.

Sanchez, Marla C., Richard E. Brown, Carrie Webber, and Gregory K. Homan. "Savings estimates for the United States Environmental Protection Agency's ENERGY STAR voluntary product labeling program." *Energy Policy* 36, no. 6 (2008): 2098-2108.

Sandel, Megan, et al. "Housing interventions and control of health-related chemical agents: A review of the evidence." *Journal of Public Health Management & Practice* 16, no. 5 (2010): S24-S33.

Sanders, Kelly T., and Michael E. Webber. "Evaluating the energy consumed for water use in the United States." *Environmental Research Letters* 7 (2012).

Sawyer, Robert F. "Vehicle emissions: Progress and challenges." *Journal of Exposure and Environmental Epidemiology* 20 (2010): 487-488.

Schueler, Thomas R. "The importance of imperviousness." *Watershed Protection Techniques* 1, no. 3 (1994): 100-111.

Schueler, Thomas R., Lisa Fraley-McNeal, and Karen Cappiella. "Is impervious cover still important? Review of recent research." *Journal of Hydrologic Engineering* 14, no. 4 (2009): 309-315.

Schweitzer, Lisa, and Jiangping Zhou. "Neighborhood air quality, respiratory health, and vulnerable populations in compact and sprawled regions." *Journal of the American Planning Association* 76, no. 3 (2010): 363-371.

Sciera, Katherine L., et al. "Impact of land disturbance on aquatic ecosystem health: Quantifying the cascade of events." *Integrated Environmental Assessment and Management* 4, no. 4 (2008): 431-442.

Shen, Li-Yin, Jian Li Hao, Vivian Wing-Yan Tam, and Hong Yao. "A checklist of assessing sustainability performance of construction projects." *Journal of Civil Engineering and Management* 13, no. 4 (2007): 273-281.

Short, John Rennie. "Metropolitan USA: Evidence from the 2010 Census." *International Journal of Population Research* 2012, (2012).

Simberloff, Daniel. "Non-natives: 141 scientists object." *Nature* 475 (2011): 36.

Sister, Chona, Jennifer Wolch, and John Wilson. "Got green? Addressing environmental justice in park provision." *Geojournal* 75, no. 3 (2010): 229-248.

Smith, W. Brad, and David Darr. *U.S. Forest Resource Facts and Historical Trends.* U.S. Forest Service. 2004. http://www.fia.fs.fed.us/library/briefings-summaries-overviews/docs/2002_ForestStats_%20FS801.pdf.

Smith, W. Brad, Patrick D. Miles, Charles H. Perry, and Scott A. Pugh. *Forest Resources of the United States, 2007.* U.S. Forest Service. 2009. http://www.nrs.fs.fed.us/pubs/7334.

Southworth, Frank, and Anthon Sonnenberg. "Set of comparable carbon footprints for highway travel in metropolitan America." *Journal of Transportation Engineering* 137, no. 6 (2011): 426-435.

Sperry, Benjamin R., Mark W. Burris, and Eric Dumbaugh. "A case study of induced trips at mixed-use developments." *Environment and Planning B: Planning and Design* 39, no. 4 (2012): 698-712.

Spielman, Derek, Barry W. Brook, and Richard Frankham. "Most species are not driven to extinction before genetic factors impact them." *Proceedings of the National Academies of Science* 101, no. 42 (2004): 15261-15264.

Sprague, Lori A., Douglas A. Harned, David W. Hall, Lisa H. Nowell, Nancy J. Bauch, and Kevin D. Richards. *Response of Stream Chemistry During Base Flow to Gradients of Urbanization in Selected Locations Across the Conterminous United States, 2002–04.* U.S. Geological Survey. 2007. http://pubs.usgs.gov/sir/2007/5083.

Stedman, Susan-Marie, and Thomas E. Dahl. *Status and Trends of Wetlands in the Coastal Watersheds of the Eastern United States 1998 to 2004.* National Oceanic and Atmospheric Administration; National Marine Fisheries Service; and U.S. Department of the Interior, Fish and Wildlife Service. 2008. http://www.fws.gov/wetlands/Documents/Status-and-Trends-of-Wetlands-in-the-Coastal-Watersheds-of-the-Eastern-United-States-1998-to-2004.pdf.

Stone, Brian. "Urban sprawl and air quality in large U.S. cities." *Journal of Environmental Management* 86, no. 4 (2008): 688-698.

Stone, Brian, Adam C. Mednick, Tracey Holloway, and Scott N. Spak. "Is compact growth good for air quality?" *Journal of the American Planning Association* 73, no. 4 (2007): 404-418.

Stone, Brian, Adam C. Mednick, Tracey Holloway, and Scott N. Spak. "Mobile source CO_2 mitigation through smart growth development and vehicle fleet hybridization." *Environmental Science & Technology* 43, no. 6 (2009): 1704-1710.

Stone, Brian, and John M. Norman. "Land use planning and surface heat island formation: A parcel-based radiation flux approach." *Atmospheric Environment* 40 (2006): 3561-3573.

Stone, Brian, Jeremy J. Hess, and Howard Frumkin. "Urban form and extreme heat events: Are sprawling cities more vulnerable to climate change than compact cities?" *Environmental Health Perspectives* 118, no. 10 (2010): 1425-1428.

Subramanian, Rajesh. "Motor vehicle traffic crashes as a leading cause of death in the United States, 2007." *Traffic Safety Facts.* National Highway Traffic Safety Administration. 2011. http://www-nrd.nhtsa.dot.gov/Pubs/811443.pdf.

Suding, Katharine N. "Toward an era of restoration in ecology: Successes, failures, and opportunities ahead." *Annual Review of Ecology, Evolution, and Systematics* 42 (2011): 465-487.

Sutton, Paul C., Sharolyn J. Anderson, Christopher D. Elvidge, Benjamin T. Tuttle, and Tilottama Ghosh. "Paving the planet: Impervious surface as proxy measure of the human ecological footprint." *Progress in Physical Geography* 33, no. 4 (2009): 510-527.

Tallamy, Douglas W., and Kimberley J. Shropshire. "Ranking lepidopteran use of native versus introduced plants." *Conservation Biology* 23, no. 4 (2009): 941-947.

Taylor, Brian D., Douglas Miller, Hiroyuki Iseki, and Camille Fink. "Nature and/or nurture? Analyzing the determinants of transit ridership." *Transportation Research Part A: Policy and Practice* 43, no. 1 (2009): 60-77.

Theobald, David M., and William H. Romme. "Expansion of the U.S. wildland-urban interface." *Landscape and Urban Planning* 83, no. 4 (2007): 340-354.

Theobald, David M., Scott J. Goetz, John B. Norman, and Patrick Jantz. "Watersheds at risk to increased impervious surface cover in the conterminous United States." *Journal of Hydrologic Engineering* 14, no. 4 (2009): 362-368.

Tiefenthaler, Liesl L., Eric D. Stein, and Kenneth C. Schiff. "Watershed and land use-based sources of trace metals in urban storm water." *Environmental Toxicology and Chemistry* 27, no. 2 (2008): 277-287.

Transportation Research Board of the National Academies. *Climate Change and Transportation: Summary of Key Information.* 2012. http://onlinepubs.trb.org/onlinepubs/circulars/ec164.pdf.

—. *Land Use and Site Design: Traveler Response to Transportation System Changes.* 2003. http://onlinepubs.trb.org/onlinepubs/tcrp/tcrp_rpt_95c15.pdf.

Trojan, Michael D. "Land use impacts on groundwater quality." *Water Encyclopedia* 5 (2005): 250-253.

Trojan, Michael D., Jennifer S. Maloney, James M. Stockinger, Erin P. Eid, and Mark J. Lahtinen. "Effects of land use on ground water quality in the Anoka Sand Plain Aquifer of Minnesota." *Ground Water* 41, no. 4 (2003): 482-492.

Tzoulas, Konstantinos, et al. "Promoting ecosystem and human health in urban areas using green infrastructure: A literature review." *Landscape and Urban Planning* 81, no. 3 (2007): 167-178.

United States Conference of Mayors. *Recycling America's Land: A National Report On Brownfields Redevelopment, Volume VII.* 2008. http://www.usmayors.org/brownfields/brownfields_bp08.pdf.

—. *Vacant and Abandoned Properties: Survey and Best Practices.* 2009. http://www.usmayors.org/pressreleases/uploads/VACANTANDABANDPROP09.pdf.

U.S. Census Bureau. *1940 Census of Population and Housing—Families.* 1943. http://www2.census.gov/prod2/decennial/documents/41272176.zip.

—. *1980 Census of Population: United States Summary.* 1983. http://www2.census.gov/prod2/decennial/documents/1980/1980censusofpopu8011uns_bw.pdf.

—. *1990 Census of Population and Housing: Population and Housing Unit Counts United States.* 1993.

—. *2010 Census Urban and Rural Classification and Urban Area Criteria.* http://www.census.gov/geo/reference/ua/urban-rural-2010.html (accessed August 6, 2012).

—. *American FactFinder.* http://factfinder2.census.gov/faces/nav/jsf/pages/index.xhtml (accessed February 4, 2013).

—. "American housing survey for the United States: 2009." *Current Housing Reports, Series H150/09.* 2010. http://www.census.gov/compendia/statab/2012/tables/12s0990.pdf.

—. *Annual Estimates of the Resident Population for the United States, Regions, States, and Puerto Rico: April 1, 2000 to July 1, 2009.* 2009. http://www.census.gov/popest/data/historical/2000s/vintage_2009/index.html.

—. *Changes in Urbanized Areas from 2000 to 2010.* 2010. http://www2.census.gov/geo/ua/PopAreaChngeUA.xls.

—. *Commuting in the United States: 2009.* 2011. http://www.census.gov/prod/2011pubs/acs-15.pdf.

—. *Current Population Survey.* 2011. http://www.census.gov/population/socdemo/hh-fam/hh6.xls.

—. *Historical Census of Housing Tables: Living Alone.* http://www.census.gov/hhes/www/housing/census/historic/livalone.html (accessed February 4, 2013).

—. *Historical Census of Housing Tables: Units in Structure.* http://www.census.gov/hhes/www/housing/census/historic/units.html (accessed January 2, 2013).

—. *Historical National Population Estimates: July 1, 1990 to July 1, 1999.* 2000. http://www.census.gov/popest/data/national/totals/pre-1980/tables/popclockest.txt.

—. *Housing Characteristics: 2010.* 2011. http://www.census.gov/prod/cen2010/briefs/c2010br-07.pdf.

—. *Means of Transportation to Work for the U.S.* http://www.census.gov/hhes/commuting/files/1990/mode6790.txt (accessed February 28, 2013).

—. *Median and Average Square Feet of Floor Area in New Single-Family Houses Completed by Location.* n.d. http://www.census.gov/const/C25Ann/sftotalmedavgsqft.pdf.

—. "Notice of final program criteria." *Federal Register*, August 24, 2011: 53029-53043.

—. *Percent Urban and Rural in 2010 by State.* 2010. http://www2.census.gov/geo/ua/PctUrbanRural_State.xls.

—. *POP Culture: 1900.* n.d. http://www.census.gov/history/pdf/History_1900.pdf.

—. *Population and Housing Unit Estimates.* http://www.census.gov/popest (accessed August 13, 2012).

—. *Population and Land Area of Urbanized Areas, for the United States: 1970 and 1960.* 1979. http://www2.census.gov/prod2/decennial/documents/31679801no108ch1.pdf.

—. "Population of urbanized areas: 1950 and 1960." *Census of Population: 1960 Supplementary Reports.* 1961.

—. *Selected Housing Characteristics: 2011 American Community Survey 1-Year Estimates.* http://factfinder2.census.gov/bkmk/table/1.0/en/ACS/11_1YR/DP04 (accessed January 2, 2013).

—. *State & County QuickFacts.* http://quickfacts.census.gov/qfd/states/00000.html (accessed February 27, 2013).

—. *United States Summary: 2000 Population and Housing Unit Counts.* 2004. http://www.census.gov/prod/cen2000/phc3-us-pt1.pdf.

—. "The urban and rural classifications." *Geographic Areas Reference Manual.* 1994. http://www.census.gov/geo/reference/garm.html.

U.S. Department of Agriculture. *Summary Report: 2007 National Resources Inventory.* 2009. http://www.nrcs.usda.gov/Internet/FSE_DOCUMENTS//stelprdb1041379.pdf.

U.S. Department of Agriculture, U.S. Forest Service. *National Report on Sustainable Forests—2010.* 2011. http://www.fs.fed.us/research/sustain/national-report.php.

—. *U.S. Forest Resource Facts and Historical Trends.* 2009. http://www.fia.fs.fed.us/library/brochures/docs/Forest%20Facts%201952-2007%20English%20rev072411.pdf.

U.S. Department of Energy. *2011 Buildings Energy Data Book.* 2012. http://buildingsdatabook.eren.doe.gov/docs%5CDataBooks%5C2011_BEDB.pdf.

—. *National Best Practices Manual for Building High Performance Schools.* 2002. http://apps1.eere.energy.gov/buildings/publications/pdfs/energysmartschools/nationalbestpracticesmanual31545.pdf.

U.S. Department of Transportation. *National Transportation Statistics.* 2012. http://www.bts.gov/publications/national_transportation_statistics/pdf/entire.pdf.

U.S. Energy Information Administration. *Annual Energy Review 2010.* 2011. http://www.eia.gov/totalenergy/data/annual/archive/038410.pdf.

—. "Table HC10.1." *2009 Residential Energy Consumption Survey Data: Total Square Footage of U.S. Homes.* 2012. http://www.eia.gov/consumption/residential/data/2009.

U.S. General Services Administration. *The Benefits and Challenges of Green Roofs on Public and Commercial Buildings.* 2011. http://www.gsa.gov/portal/getMediaData?mediaId=158783.

U.S. Geological Survey. *Ground-Water Depletion Across the Nation.* 2003. http://pubs.usgs.gov/fs/fs-103-03.

—. *National Assessment of Coastal Vulnerability to Sea-Level Rise.* http://woodshole.er.usgs.gov/project-pages/cvi (accessed April 3, 2013).

—. *Water Use in the United States.* http://water.usgs.gov/watuse (accessed November 1, 2012).

U.S. Global Change Research Program. *Global Climate Change Impacts in the United States.* Cambridge University Press. 2009. http://nca2009.globalchange.gov/download-report.

U.S. Government Accountability Office. *Comprehensive Asset Management Has Potential to Help Utilities Better Identify Needs and Plan Future Investments.* 2004. http://www.gao.gov/products/GAO-04-461.

Vaccaro, J.J., and T.D. Olsen. *Estimates of Ground-Water Recharge to the Yakima River Basin Aquifer System, Washington, for Predevelopment and Current Land-Use and Land-Cover Conditions.* U.S. Geological Survey. 2007. http://pubs.usgs.gov/sir/2007/5007/index.html.

Vicino, Thomas J. "The political history of a postwar suburban society revisited." *History Compass* 6, no. 1 (2008): 364-388.

Vincent, Grayson K., and Victoria A. Velkoff. *The Older Population of the United States: 2010 to 2050.* U.S. Census Bureau. 2010. http://www.census.gov/prod/2010pubs/p25-1138.pdf.

Walsh, Christopher J., Allison H. Roy, Jack W. Feminella, Peter D. Cottingham, Peter M. Groffman, and Raymond P. Morgan II. "The urban stream syndrome: Current knowledge and the search for a cure." *Journal of the North American Benthological Society* 24, no. 3 (2005): 706-723.

Wang, Sheng-Wei, Mohammed A. Majeed, Pei-Ling Chu, and Hui-Chih Lin. "Characterizing relationships between personal exposures to VOCs and socioeconomic, demographic, behavioral variables." *Atmospheric Environment* 43, no. 14 (2009): 2296-2302.

Wang, Youfa, and May A. Beydoun. "The obesity epidemic in the United States—gender, age, socioeconomic, racial/ethnic, and geographic characteristics: A systematic review and meta-regression analysis." *Epidemiologic Reviews* 29, no. 1 (2007): 6-28.

Wear, David N. *Forecasts of County-Level Land Uses Under Three Future Scenarios: A Technical Document Supporting the Forest Service 2010 RPA Assessment.* U.S. Department of Agriculture, U.S. Forest Service. 2011. http://www.srs.fs.usda.gov/pubs/gtr/gtr_srs141.pdf.

Wenger, Seth J., James T. Peterson, Mary C. Freeman, Byron J. Freeman, and D. David Homans. "Stream fish occurrence in response to impervious cover, historic land use, and hydrogeomorphic factors." *Canadian Journal of Fisheries and Aquatic Sciences* 65, no. 7 (2008): 1250-1264.

Weschler, C.J. "Chemistry in indoor environments: 20 years of research." *Indoor Air* 21, no. 3 (2011): 205-218.

Weschler, Charles J. "Changes in indoor pollutants since the 1950s." *Atmospheric Environment* 43, no. 1 (2009): 153-169.

Wilcove, David S., David Rothstein, Jason Dubow, Ali Phillips, and Elizabeth Losos. "Quantifying threats to imperiled species in the United States." *BioScience* 48, no. 8 (1998): 607-615.

Woltemade, Christopher J. "Impact of residential soil disturbance on infiltration rate and stormwater runoff." *Journal of the American Water Resources Association* 46, no. 4 (2010): 700-711.

Xian, George, Collin Homer, Jon Dewitz, Joyce Fry, Nazmul Hossain, and James Wickham. "Change of impervious surface area between 2001 and 2006 in the conterminous United States." *Photogrammetric Engineering & Remote Sensing* 77, no. 8 (2011): 758-762.

Yip, Fuyuen Y., Jeffrey N. Pearcy, Paul L. Garbe, and Benedict I. Truman. "Unhealthy air quality—United States, 2006-2009." *Morbidity and Mortality Weekly Report* 60 (2011): 28-32.

Younger, Margalit, Heather R. Morrow-Almeida, Stephen M. Vindigni, and Andrew L. Dannenberg. "The built environment, climate change, and health: Opportunities for co-benefits." *American Journal of Preventive Medicine* 35, no. 5 (2008): 517-526.

Yuan, Fei. "Urban growth monitoring and projection using remote sensing and geographic information systems: A case study in the Twin Cities Metropolitan Area, Minnesota." *Geocarto International* 25, no. 3 (2010): 213-230.

Zogorski, John S., et al. *Volatile Organic Compounds in the Nation's Ground Water and Drinking-Water Supply Wells.* U.S. Geological Survey. 2006. http://pubs.usgs.gov/fs/2006/3048.